THE WEAPON ON THE WALL

Rethinking Psychological Warfare

THE WEAPON ON THE WALL

Rethinking Psychological Warfare

BY MURRAY DYER

The Johns Hopkins Press: Baltimore

Distributed in Great Britain by Oxford University Press, London

Printed in the United States of America by Franklin Printing Co., Primos, Pa.

Library of Congress Catalog Card Number 59-14234

This book has been brought to publication with the assistance of a grant from The Ford Foundation.

for E A J

" . . . in close fight a champion grim,

In camps a leader sage."

MARMION

Dim drums throbbing, in the hills half heard,
Where only on a nameless throne
a crownless prince has stirred,
Where, risen from a doubtful seat
and half-attainted stall,
The last knight of Europe
takes weapons from the wall,
The last and lingering troubadour
to whom the bird has sung,
That once went singing southward
when all the world was young.

G. K. CHESTERTON

Foreword

The study reported in this book resulted from one of those extraordinary and fortunate accidents for which we are always grateful. In this case the event was the meeting of an excellent mind, the mind of a superb research man with extensive experience in four of the major cultures of the world, the United States, England, India and Japan; skilled in the practical arts of communications because of long experience in the Associated Press and other newspapers, as an author, and in radio communications with NBC and CBS; and trained in world-wide application of political communications through his experience in the Office of War Information during World War II.

Mr. Murray Dyer joined the Operations Research Office fortunately and fortuitously at the time when it was conducting on behalf of the Army one of the greatest research programs that has ever been conducted in psychological warfare outside of the Soviet Union. It was also at a time when some of the new research results could be tested in the war in Korea, and when new and critical information on political communications could be obtained from that well fought but sadly mismanaged war. On the basis of his matured experience, reinforced by ORO research results, he then had an opportunity during his later work for ORO in the United States, in England, and in Europe to discuss anew the problems of political communications with the scholars and leaders of the free world. Finally, in connection with classified studies, it was ORO's good fortune, as well as that of Mr. Dyer, to be given free access to all of the great mass of highly classified papers at the top levels of the United States government related to psychological warfare. These papers, not yet officially synthesized, discuss the policies and decisions of past times—policies and decisions which led in some cases to triumphs and in some cases to failures. Direct use of this classified information was not permitted to Mr. Dyer in this book, but in no sense is it contradictory to the lessons or conclusions of this study.

In all of his experience, Mr. Dyer was a professional, from his several years as a teacher in the Imperial Japanese Naval College, as a newspaper man, as a radio commentator, as an author, as a practitioner in psychological warfare, and as a research man. For this study the opportunity was unique, the man was just right. This is a professional study.

I agree with Murray Dyer. We need to take down the weapon from the wall. We need to weld together—in the free world—words and deeds. Our deeds are the nursing mother of our political communications, and I mean our communications to our own people, as well as to all of the people of the world, free and slave alike. Without welding our deeds and words together, we suck air. Random, spur of the moment, emotional, or unplanned cold war deeds by the United States, which are unrelated to political communications, have only the nourishment of air.

This well considered study has been long needed. I hope it will influence the free world to take down and use The Weapon on the Wall.

Ellis A. Johnson, DIRECTOR
OPERATIONS RESEARCH OFFICE
THE JOHNS HOPKINS UNIVERSITY

Preface

As one drives up a long and winding road into mountains, one gets from time to time a long, clear view of where one has been, and again an enclosed and narrow view of where one is going, and sometimes a glimpse of the distant rise with the ribbon of road on it where one will be after a time. The view of any of the great issues that confronted the United States in the past generation, and may confront it for decades yet in the future, has some resemblance to this. When we were quite young, we knew of books with titles such as *America Comes of Age*. As we have grown older, we have found the coming of age a long drawn-out experience. The top of the hill has receded as we climb. The pinnacles we saw, and attained, were only foothills.

One of the things that we have found it necessary to accept is propaganda or psychological warfare by our government. Not so long ago we thought that the government should have nothing to do with such matters. But in World War I we accepted propaganda as a necessary kind of activity in wartime. More readily we accepted it again in World War II. Psychological warfare, sometime between 1939 and 1945, became one of those things of which many of us were rather fond. It was a mark of our superiority over the isolationism of a few years earlier. It marked the acceptance of the necessity for American engagement in the world. To recognize that psychological warfare could be important and necessary was a badge of sophistication. We have accepted something of this general sort as necessary in the continuing time of stress that we tag as cold war, but we have not been quite so confident as in simple war.

Mr. Dyer employs a fresh term, *political communication,* instead of *psychological warfare* or *propaganda.* Had this book been written just a few years sooner, this change of term might not have seemed necessary. But it is now six years since the truce in Korea, and eight years since the truce negotiations started. It is five years since the truce in Indo-China. We have no war, in the familiar sense, no front line rippling with shellbursts and machine-gun clatter. Yet we still have a situation in which the United States finds it necessary to maintain an agency for what is essentially propaganda.

The concept of propaganda or psychological warfare was shaped in 1917 and again in 1940 by the fact of transition from assumed peace to assumed war. It has to be shaped now, presumably, by the assumption of United States engagement in a struggle of indefinite duration that is neither war nor peace as we used to know them. To assert that this makes a big difference may be puzzling. Yet in such practical connections as roles and missions for parts of the United States government, it carries simple consequences. The mission of psychological warfare activity in war was to help in winning the war. The appropriate roles of agencies could be related to their roles in the war itself. War, and victory, were wholly familiar traditional concepts and supposedly needed no explanations. Once established as a premise, war was accepted by all concerned as better to win. This did not solve all problems about the conduct of propaganda, but it solved a number of them that are very difficult to solve without that premise. We cannot go again from a state of peace, with no propaganda organization, to a state of war, in which we quickly organize one. If we go to war again, we go with an old organization already in the government.

This does not make it easy to say how things ought to be done. It does imply that we cannot simply open the old book and observe the old lessons. Too many of the old lessons will simply not be pertinent.

The changes in the role of the United States, from that of an innocent bystander in world politics to that of a major participant and protagonist, are reflected in more ways than just in the need for an information service. We are keeping up a program of military aid to foreign nations at a time when no specific war is looming in the immediate future. We are keeping up our own armaments on a scale that is supposedly adequate for full-scale engagement at the very start of a war, not counting on allies to buy us time, as formerly, to recognize our interests and to arm ourselves. We are concerned with a structure of alliances that greatly surpasses any older one in geographic scope and complexity. We are concerned about the problems of population and productivity and welfare, and consequent political stability or lack of it, in areas far greater in population than our own country.

The exact goal of the United States in all this is no more clear than the distant view of the peak we are climbing, which we cannot always see at all and sometimes only through broken clouds. Any sharp statement of the United States role surely is an invitation to debate, even among good friends. But what is debatable falls around the edges and along the seams of a bigger cloth that is undebatable. The United States is something like one party or the leader of one party in the world's political system. It can no more be silent to the peoples of the world than a party in American politics can forsake the claim to equal time on TV and radio. The world is engaged in mighty decisions, by war or by peace or by some competitive and bitter conflict that is not war. The decisions to be reached will govern how much of the principles embodied in the American system, or how much of the Communist system, will characterize the whole world in the next century. If this were to be only by debate, an agency of the United States government engaged in the debate would be as necessary as a speaker's bureau is necessary for a party in a campaign year. The cold war is a political campaign of sorts, on the scale of the world, without a world government, but still a campaign. No one really proposes that the United States should not run.

The budget for the United States foreign information program faces trouble in Congress year after year, but the question is on the plane of how much to grant and how to organize; it is no longer a question of whether to keep it up or stop it. But the idea does seem to hang on that the matter is related to war in some way and that the fact of war, of a sort, is essential as a justification of using such means. We have a familiar and usable concept of propaganda or psychological warfare related to war, and we have the traditional idea that it should be excluded in peace. Neither of these ideas fits comfortably in the current circumstances, and we have not formulated any new concept free of the old and now false dichotomy. The awkwardness with which we handle the problem is a symptom of the awkwardness of thought based on inconvenient concepts.

The degree to which war has served as a premise for American concepts about propaganda was not much observed until lately. Granted, we knew ten or twenty years ago that a war did serve to settle the issue as to whether the United States could engage in such activities at all. Still, we scarcely noted the great degree to which this premise, war, made many problems easy to solve. What is the purpose or object of propaganda? That is easily answered when the answer is "Can the Kaiser" or "Beat Hitler." What control over what parts of a propaganda program should be under civilian and what under military authority? The answer is at least made easier by the existence and stature and claims of field commands having responsibility for vast armies and fleets and air forces engaged in gigantic battles.

The Creel Committee and the Office of War Information both had their tasks simplified by the premise that there was a war to win. This simplified matters in more ways than one. It settled most arguments about the justification of means. It was familiar and passed unchallenged in argument. And it served as a unifying doctrine for persons of both parties and all shades of opinion on other matters. It would have been extremely difficult to establish a usable doctrine for common action on any more sophisticated premises.

Our old wars had another characteristic that was important. They lasted a few years, years of effort to organize all the national resources into the means required for victory. Also, if we go to war again, we cannot expect to have years of time in which to mobilize the national resources. Either the time or the resources themselves to a major degree may be denied us. The assumption cannot be that we have several years in which to make professionals out of amateurs. The United States Information Agency and other agencies engaged in "presenting" the American case to the world in the last five years, or in "co-ordinating" the information output and other activities, have had to plow new ground, and in every way in which the premise of a present war makes such things easier, they have lacked that premise and things have not been so easy.

What is shown in this book is the record of cloudiness in

our national councils on questions about the purposes and techniques and organization of psychological warfare or of the foreign information function. If even to have such an organization is to accept the premise of war, then military agencies might exert a greater weight in its control and carry more of the burden of execution of a program. If the end is to help win one war, the means that are justified may be different in a number of respects than they would be if the end is to win an indefinite world political struggle without war. Differences of basic premises, and the difference of derived conclusions that spring from them, have confused our policy-makers and organizational moves for the last fifteen years, more or less. Are the old concepts of propaganda and psychological warfare adequate? Is the assumption that such things arise with war and disappear in peace adequate? Is the assumption that because we need them now we are committed to the notion of war adequate? Must we pretend that this is war in order to keep up an activity that we once conceived of as proper only to war?

The fresh insight that this book brings to the subject goes beyond any matter of detail. It concerns the whole nature of the thing. The task of the United States government to present to the whole world the rationale of American principles and actions should, in short, not be a mere adjunct and auxiliary means to winning some particular war. The situation has evolved beyond that. War, if it occurs, will be an incident in a larger situation requiring political communication, and winning it cannot be a simple and sufficient end in itself.

JUNE 15, 1959

George S. Pettee
ASSISTANT DIRECTOR
OPERATIONS RESEARCH OFFICE
THE JOHNS HOPKINS UNIVERSITY

Acknowledgments

The manuscript of this book has a long history of tribulation. It has been under review in various government quarters for sixteen months. The fact that it is now being published is due, in no small measure, to the integrity of the United States Army, and it is reassuring and encouraging that, at a time when the problems of censorship are omnipresent, it was the Army, on its own initiative, that finally authorized publication of independent thinking for which it accepts no responsibility and with which it undoubtedly does not entirely agree.

I also wish to acknowledge my indebtedness to Dr. Edward P. Lilly. The fact that I have been able to write authoritatively in Chapter IV about certain of our experiences with psychological warfare in World War II is due to his work and interest in this field. He has not yet published, but in several instances I have drawn upon his knowledge. I should have liked to record more specifically this indebtedness. I can hope only that it will not be long before he, himself, makes his own contribution in writing to our knowledge of psychological warfare in World War II. In my estimation it would prove to be the most important single history of psychological warfare yet written.

Finally, for the concept of revolution that I have taken in this book I am indebted to Dr. George S. Pettee and his writings on the subject.

There are other indebtednesses too numerous to mention.

AUGUST, 1959 *Murray Dyer*

Contents

I The Role of the Weapon

There are two sets of dates in the twentieth century that history may well mark as either the beginning of the end of the Western heritage, or else the beginning of its "finest hour."

In this moment of truth it is not the engines of destruction that will overwhelm us, nor is it the physical achievements possible to science and industry that will liberate us. It will be the cast and tenor of our minds, the temper and resiliency of our thinking, the worth and value of our ideas that will determine our destiny. We have the knowledge, and it is growing at astounding speed. It is how we propose to use it that will put us to the test. And it is the two sets of dates in this century that so clearly tell us that the time has come for full mobilization of all our resources if we intend to preserve and develop the freedoms of the human spirit.

Curiously, both sets of dates are paired, as if to drive the lessons home and give us no excuse for scamping our homework. The first two dates are August 6 and 9, 1945, at which time we dropped atomic bombs on Hiroshima and Nagasaki and, in doing so, announced the opening of the nuclear age. These weapons were first used by a free power, instinct with the heritage of the West, animate with the traditions of ancient Greece, openly acknowledging in its cathedrals the omnipotence of a God who requires that man love his neighbor. The point to notice is that the nation that first used these weapons, in whose very bones are bred the value and worth of ideas, at the time it made its inevitable yet terrible decision ignored in every important respect the psychological nature and attributes of its action. As Thomson has said:

> It was hardly a surprise to the propagandist that he should have had no opportunity whatsoever to help central leaders estimate the effects on world opinion of the atomic bomb. Even in such a revolutionary case, the leaders relied on their own rule-of-thumb estimates, which, however reliable for judging United States opinion, were certainly not adequate to cover the complexities of world reaction. In consequence, our propagandists faced the greatest strategic crisis of our history with no forewarning and no chance to plan what should be done to maximize favorable effects, and to offset bad ones; nor was there any plan offered by the leaders.[1]

In the interval since then the patent fact that we are in a war of ideas has been similarly neglected. At times much play of lip service has been given to our Western heritage and the worth of ideas in our way of life. But in terms of substantive action, designed to make clear and apparent to all free men that this heritage is a living force to be reckoned with, much remains to be done.

Like a thunderclap on a summer's afternoon came the second pair of dates—October 4 and November 3, 1957. On these days Russia launched Sputnik I and Sputnik II, respectively. And in launching them the power that denies the dignity of men, that has suppressed the aspirations of

[1] Charles A. H. Thomson, *Overseas Information Service of the United States Government* (Washington, D. C.: The Brookings Institution, 1948), pp. 364-65.

the human spirit, that has trampled underfoot the finest flowerings of the human mind, showed unmistakeably that it recognized the nature of the link between action and psychological consequence.

With malice, forethought, integrated governmental action, and skill it marshaled its case before the world and proclaimed a right to leadership on the ground that its knowledge was superior. On the edge of the nuclear age, ushered in by the free world, it opened the doors to the space age and combined both prospects in a series of calculated and psychologically correct assessments of the impact upon the minds of men.[2]

The significant aspect of these two memorable sets of dates is this: we were unprepared for, and evidently unmindful of, what men might think about the use to which we put knowledge that we acquired, while the Russians were not only prepared and mindful, but took pains to impress upon the world the meaning they intended this knowledge to have.

With this second lesson staring us in the face, the immediate response of our national leadership was uncomfortably reminiscent of our own attitude at the time of Hiroshima and Nagasaki. Nothing really was different from what it had been yesterday, we were told.[3] Literally, true enough. But in terms of ideas, once more men had broken through into a field full of the promise of knowledge different both in kind and degree. And once more, as in August of 1945, no yesterday held comfort, security, or guidance. A new world was being born, requiring new ideas in order to mature. Willy-nilly, men will accept and act on these ideas, and the world we know will respond to the devices of men's minds. The important point is the premises on which these ideas are founded. It is here that we have overlooked the

2 See files of the *New York Times* for this period.
3 *Ibid.*

psychological factors while giving weight to the political, the economic, and the military aspects. And it is here that the Russians have shown, with force and certainty, the role of the minds of men in the struggle that is, and will be until resolved, the controlling element in terms of which our decisions have to be cast.

The Nature of the Present Conflict

The most significant element in the nature of the present conflict is the element of revolution. Equally significant, the dynamic, planned, and long-range leadership of this revolution is in Russian hands. The United States, and the free world, are on the defensive. Our policy as a nation has been reactive; that is, we have waited to act until forced by events to do so, rather than anticipating events and being ready to take action when they arrived.

The Truman Doctrine, the Marshall Plan, the Korean war, even as recent an event as the NATO conference at Paris in December, 1957, all document this position. As leaders of the free world, we had to take certain action, but not until situations had become critical did we react. Then we "muddled through," brilliantly, as in the instance of the Marshall Plan, but nevertheless without prior planning, without having first mobilized our resources and determined the best course to follow.

There are, of course, many reasons for this. Our role as a world leader is hardly more than a decade old. Because of the increasing speed of modern technology and the scope of its impact on the world, we have had to learn in action and under pressure. There has been relatively little time for thought and reflection. Men and women brought up under a concept of our nation as a state separate and distinct from the old world, freed of its shibboleths and passions, have been forced by the logic of the revolution now taking place to assume responsibilities reaching literally to the ends of the earth in order to protect and advance the way of life we cherish.

This revolution is not *within* states making up one world. It is a revolution throughout that world, and its manifestations are apparent in our technological progress, our economic problems, and the machinery we have developed for our communications. Put in perhaps its simplest form, the fact of nuclear war is requiring us to re-examine and rethink such concepts as sovereignty, independence, nationalism, and war itself. We are no longer entitled to regard the independence of a state as an end in itself when that independence involves the prospect of force for its establishment or maintenance and when that force is nuclear in application. The possibility of survival under such circumstances is altering ancient habits of thought, but habits of thought are slow to change, and they change at different speeds according to the acuteness with which cause and effect are perceived and understood.

Essentially we are in a world conflict of ideas, and the resolution of these conflicting ideas may not be through force—the traditional accompaniment of revolution—unless we choose to risk our heritage. All this is known and has been said. The distinctive element of the problem is the successful accomplishment of this revolution, together with the preservation of freedom, without recourse to force as the final arbiter. Under such circumstances we are required, if we would win, to fall back on our wits. This is precisely what the Russians have done. They have calculated, soberly and with nicety, the lengths to which they may go before invoking the use of force against them on a global scale, they have used diplomacy and economic sanction, they have become masters of intrigue and sedition, but over and above all this they have perverted ideas and so manipulated them as to give plausibility to their objectives. And, as in the military, diplomatic, and economic fields, our defense has been yet once more reactive.

We have wasted our time arguing with Russia, attempting to deny the premises for her ideas, seeking to preserve a status quo. In other words, we ourselves have not yet fully perceived the nature of the revolution that confronts us. Never again shall we see yesterday. And just as surely we cannot remain detached from this revolution and expect that tomorrow we shall be unharmed. The critical question is not the preservation of outmoded social forms, but the evolution of new forms, still preserving those values and rewards that are the heritage of Western civilization. This permits a turning of the tables and, properly conceived, puts us in the role of leading such a revolution and bringing it to a successful conclusion.

It is in point to note here that this revolution is not solely dependent on a viable, active, and malevolent Soviet Union. Were Russia to disintegrate, we should still be faced with a revolution in thought and ideas affecting the minds of the people of the world. And this revolution would still demand positive leadership from us if we wished to preserve our Western heritage. It is in terms of Russia as the chief antagonist that we must presently take action. But it is well to remember that a disintegrated Russia would not end the revolution; it would only alter the terms on which the conflict would have to be prosecuted.

In terms of Hiroshima and Sputnik, and their implications, it is perhaps permissible to assert that this is the greatest challenge a nation has yet had to face. Self-evidently, we may fail. Equally self-evident is the fact that if we win without war we shall, as a people, have risen to a new height in the long development of the human spirit.

A challenge of this nature cannot be met by anything less than mobilization of a nation's will and spirit. Such a mobilization is inclusive. There must be military preparedness, economic foresight and planning, diplomatic initiative and flexibility. This, the evidence of our behavior since the close of World War II suggests, we understand and support. But if we are to succeed, there must be mobilization also of the forces inherent in a society that has deep political and ethical convictions about its way of life.

These forces are not susceptible of precise measurement. They are nevertheless powerful, and the extent and range of their power is recorded in the history of man.

The weapon primarily concerned with affecting these forces has been called psychological warfare. In the hands of our enemies it has met with outstanding success, so much so that we have felt impelled to employ it ourselves. It has been formally recognized by government as an important instrument of state. Yet the fact remains that this activity, as conducted by our own society, has no great achievement to its credit.

The seriousness of the stakes that are at issue in the present conflict makes it worth while to examine this weapon and attempt to determine the manner of its use, its boundaries, and its capabilities for serving the nation at this time.

In the first place, it is regarded as a weapon for affecting the minds of men. In the second, we are employing it, officially, in our dealings with the enemy, with neutrals, and with our friends. As a recognized instrument of government, for the use of which we can be held responsible by civilized opinion, it is required to rest on certain premises, and it ought to function under certain operating principles.

The weapon itself is not new. It has been with us from the beginnings of man. What is new about it is its employment on an organized scale by governments and in the framework of tremendous advances in technological methods of mass communication. In this context its history dates back no further than World War I, and this is the only context that is of primary concern in any examination of its role in the modern world.

In this frame of reference it is significant to notice that the history of psychological warfare is closely bound to the history of war itself. Activities in this field have been regarded as primarily military in nature, and their incorporation as an element in the weapons systems of states has given rise to its name. Psychological warfare, however, is now being employed on a far wider scale. It is an integral component of Soviet strategy for the establishment of its own world hegemony. It is a component of our own defensive strategy. It is still called, by us, psychological warfare.

The Inadequacy of the Name Psychological Warfare *and the Difficulties of Definition*

The most cursory examination is sufficient to determine the inaccuracy of this term. We are not at liberty to wage warfare against our friends or against neutrals. We may wage warfare only against our enemies. Furthermore, while, generically, we term this activity psychological warfare, at other times and among other peoples it has been given different names. It has been called political warfare, propaganda, even information.

A good definition of the activity, of course, ought to permit us to put an accurate name to it. So far it has not been found possible to define psychological warfare so that its terms of reference are clear and its boundaries determined. Considerable time and effort have been spent in trying to write definitions. Demonstrably, however, when these are applied, they are either so incomplete that they limit the activity in scope or else so general that other activities may be included. The fact is that this activity encompasses the range and sweep of the mind, and its only boundary is the universe. This does not preclude an understanding of its nature.

> In aesthetics, politics, psychology, sociology, and so forth, the stage of systematic symbolization with its fixed and unalterable definitions has not been reached. . . . The most highly systematized sciences are those which deal with the simplest aspects of nature. The more difficult and to many people, naturally, the more attractive subjects are still in a stage in which it is an open question which symbolization is most desirable.[4]

What has been called psychological warfare is indubitably in this stage in which the best symbolization is an open question. The inclusive nature of the activity, on the other hand, can be distinguished. One of the best descriptions of it has been given by a Senate committee:

> It should be borne in mind that the opinion of the world regarding the United States and its objectives is being affected continuously by news developments in the United States. In that news, the official efforts of the International Information Administration[5] play but a relatively small role. Infinitely larger is the role played by Congress, by the executive branch, the judicial branch, and by the organizations and individuals on the American scene acting on issues which might once have been considered "domestic" but which are now of deep international significance. A single action by an American organization, a speech by a Member of Congress or by an official of the executive branch, or an Executive order, administrative regulation or law, can produce—in and of itself, an impact throughout the world of vast and enduring proportion. . . . As chief of mission in a country, the United States Ambassador is in a particularly potent position to contribute formally and informally to the objectives of the information program. Wholly aside from the ordinary type of media releases, a well-directed impromptu comment of the Ambassador, a meaningful gesture, can cause

[4] C. K. Ogden and I. A. Richards, *The Meaning of Meaning* (10th ed.; London: Routledge and Kegan Paul, Ltd., 1949), pp. 121-22.

[5] Now the United States Information Agency.

> a significant impression throughout the length and breadth of the country to which he is accredited.[6]

An activity that is composed of so many elements clearly is difficult of precise definition. It is possible, however, to discuss its nature and range and to determine the reasons supporting its position as an instrument of state.

The Logic for Its Effectiveness

It is important first to consider the evidence for its effectiveness and the grounds on which this evidence rests. If its effectiveness is in doubt, and that doubt is legitimate, then clearly it must be employed with circumspection, if at all.

It is quite clear that there is considerable doubt about the effectiveness of what has been called psychological warfare. It seems equally clear, if the evidence is studied, that this doubt cannot be sustained. Emphatically, and beyond dispute, the instrument *can be effective.*

It is impossible, yet, to measure this effectiveness in terms of numbers. No one can say the activity as a whole, or specific parts of it, have been 80 per cent effective, or 50 per cent effective, or wholly effective. Such evidence as we possess on which to base a judgment—and it is now considerable—is chiefly qualitative.

First, there is the exhaustive analysis of our overseas information programs undertaken by the Senate Committee on Foreign Relations. This was the work of many months and involved detailed hearings. The verdict of this committee is unreserved.

> Recognition of the importance of the information program is general and is characteristic even of those witnesses who were highly critical of certain aspects of the

6 Report of the Committee on Foreign Relations, Overseas Information Programs of the United States (Hickenlooper Report), Senate Report No. 406, Eighty-third Congress, First Session (1953), p. 32.

program. The committee is unanimous in endorsing the continuance and the strengthening of this program.[7]

It may be assumed that such a committee, aware of the history of Congressional hostility to the concept of propaganda, would be unlikely to endorse a program unless it had become convinced of its effectiveness.

At virtually the same time that the Senate committee was investigating our overseas information program, a similar investigation was under way in Great Britain. In October, 1952, the British government of the day appointed an Independent Committee of Enquiry into the Overseas Information Services, under the chairmanship of the Earl of Drogheda, P.C., K.C.M.G. It was known as the Drogheda Committee. In issuing its report it observed:

> We have come to the following general conclusions:
>
> *First.* The Overseas Information Services play an important and indeed essential role in support of our Foreign, Commonwealth and Colonial policies.
>
> *Second.* This work should be done well, continuously and on an adequate scale.
>
> *Third.* If all these requirements are to be met more money must be spent on the Overseas Information Services.
>
> *Fourth.* Changes are required in the pattern of the work in order to bring it into line with our political, strategic and commercial needs.
>
> These conclusions have been forced upon us by sheer weight of evidence. . . . we have been left with no doubt that propaganda can pay a substantial dividend. . . .[8]

A verdict so similar in nature by two such committees is not lightly to be set aside. But this is by no means the sum of the evidence. The verdict is massively supported by the combined conclusions of responsible practitioners with ex-

[7] *Ibid.*, p. 31.

[8] Summary of the Report of the Independent Committee of Enquiry into the Overseas Information Services, April, 1954, Miscellaneous No. 12 (1954), Her Majesty's Stationery Office, London, pp. 3-4, 7.

perience in the field. Their individual verdicts coincide very closely with those of the two committees.

Again, this is not the sum of the evidence. The field of what has been called psychological warfare is one requiring pre-eminently the skills of social scientists, and many have acquired familiarity with it. And again the summed verdict supports the conclusion of the committees.

Still the evidence is not complete. There are the informed judgments of authority to be taken into account. The opinions, for example, of such men as Henry L. Stimson, Secretary of War, General Dwight D. Eisenhower, Supreme Commander of the Allied Expeditionary Forces, and Sir Winston Churchill—men who themselves are not usually considered practitioners of psychological warfare (although it is in point to note that both Sir Winston Churchill and Colonel Stimson have effectively employed it, Colonel Stimson notably with his historic letter to Senator Borah), but who have been in positions where their judgment carries weight. The evidence of such men also supports the Senate and Drogheda committees, although here it has to be observed that there are also men of authority who do not support the activity. In this instance the weight of the collective individual judgments of such authority must be balanced against those in opposition. It is the writer's opinion that this balance falls decidedly in favor of a verdict of effectiveness.

Finally, there is the evidence of the press. At first there was unquestioned hostility on the part of the press of this country to psychological warfare. Nevertheless, it is contended, the daily files of our newspapers provide significant support for the conclusion that psychological warfare is having a demonstrable effect.[9]

[9] In the course of preparing a detailed study on the subject of psychological warfare for governmental use, the files of the *New York Times*, daily and Sunday, were examined for the period February 1, 1953-February 1, 1955. This conclusion is based on this examination.

There are two further major sources of evidence that are not available to the general public. These are the classified bodies of writings on the subject of psychological warfare formed by the research studies undertaken by the Operations Research Office of The Johns Hopkins University and the RAND Corporation. Both are agencies working for the Army and the Air Force respectively. In general, therefore, much of their output is classified; less so with RAND, however, than with ORO. Both agencies have been actively engaged in research on psychological warfare. Both agencies can be recorded, on the basis of their published papers, as actively supporting, on the basis of evidence, the position that psychological warfare can be a major and important instrument.[10]

It is important to record that negative strictures against this instrument, on the part of those with professional or practical competence, are difficult to find. It will be remembered that the Senate committee itself observed in its report: "Recognition of the importance of the information program . . . is characteristic *even of those witnesses who were highly critical* of certain aspects. . . ." (Italics supplied.)

Before going further, the frame of reference of the activity itself ought to be made clear. The preceding pages surely indicate that what we have called psychological warfare is surrounded by considerable confusion as to what, in fact, falls into the domain of this activity. The Senate and Drogheda committees investigated overseas information programs. Social scientists and practitioners have written, variously, about psychological warfare, propaganda, information, and political warfare.

It is quite clear that when the body of literature now

[10] The published papers of both agencies have been available to, and read by, the writer. Unfortunately, the statement above must stand without documentation. It can be said, however, that the directors of both agencies have seen it before publication and agree with it.

available on this subject is viewed as a whole, the referent of the discussion is generally understood by those whose work is in this field, whatever name it may be given. It seems equally clear, however, that confusion nevertheless does surround the activity.

It is self-evident that the term *psychological warfare* carries false connotations. It is equally self-evident that the activity is much more than merely an overseas information program. The pertinent question to ask is: what do we desire to see accomplished by this instrument?

There are two alternative answers that merit consideration. Either we desire to convert peoples outside our own borders to an acceptance of our democratic way of life, and therefore to accept our national aims and objectives, or else we desire from them what Hans Speier has called "relevant political action."

Conversion, it is suggested, is an impossible task and, furthermore, one that ought to be repugnant. To what brand of democracy are we to convert? Shall we encourage a Democratic India and a Republican Middle East? The connotation of conversion in addition calls up a concept of the missionary proselytizing for his denomination, and a nation as heterogeneous as the United States will surely experience great difficulty finding professional propagandists who will be given solid majority support by its own public opinion for the arguments and emotions they must draw on if any attempt is made to convert other peoples.

Relevant political action offers a more realistic approach, which is consonant with what Ralph Block has called "our Lockean doctrine and . . . our respect for the right of human beings to think for themselves." In the revolutionary conflict of our day it is imperative, if we are to win, that men perceive the durability of the forces motivating the world struggle and become sufficiently aroused to support the disciplined policies necessary to victory.

Much of the record of what we have called psychological warfare shows us to have been "talking past" our audiences. Knowledge of each other's cultures has not been sufficient to permit real communication. We may not hope to win wavering allegiances as long as this continues. In addition to the will and support of our own people there must be the ability to translate this will and support into meaningful terms for those yet undecided on the nature of the present struggle or those whose courage and will need bolstering.

This calls for political knowledge and political leadership of a high order for "Only he is able to follow the course of events intelligently who comprehends the structural alignment which underlies and makes possible a given historical situation and event."[11]

This present situation is revolutionary in terms of a world order and not of states within that world. It is the rapid progression of knowledge in almost all fields that structures this revolutionary situation, and it is technology that is, in large part, responsible for the alignment of the two chief protagonists in the struggle.

If we are to win without a third world war, it must be because the nature of this alignment is clearly understood and because we put to efficient use this same knowledge and technology. In our own case this calls for a dynamic concept of what we have been calling psychological warfare, a concept that takes into account the political realities of our era, adjusts to them in terms of the deep aspirations of all men who respect freedom, and, at the same time, demonstrates that those principles we proclaimed at our founding will not be debased for mere expediency. In other words, it must be evident that our national character possesses stability.

This is much more than psychological warfare. And, from

11 Karl Mannheim, *Ideology and Utopia* (London: Routledge and Kegan Paul, Ltd., first publication in England, 1936), p. 157.

the record, it is clear that we intend our activities in the area under discussion to be more than psychological warfare. We are demonstrably searching for means of "dynamic intellectual mediation . . . as one element in the total situation," but, as Mannheim points out, "it makes a good deal of difference whether this presupposition influences one's point of view unconsciously and naively . . . or whether it appears only after everything of which we can become aware and which we already know has entered into our deliberations."[12]

This sort of mediation in our time must be primarily political in nature. We are lost if we approach the problem with missionary zeal, seeking converts to our way of life. Our way of life is open to different interpretations even among ourselves. But persuasion of the values inherent in the philosophy from which that way of life developed is, surely, a political task in the widest sense of the word political.

Political Communication *More Adequately Describes the Role of the Instrument*

If this is granted, then what we have been thinking of as psychological warfare falls, more properly, into the domain of political communication. This concept, it is suggested, may provide a key to the manner in which we should undertake the problem of seizing the initiative and offering a positive program that will counterbalance and eventually destroy, through knowledge, the philosophy proclaimed by the U.S.S.R. In essence we pass from reactive defense to planned, integrated, mobilized offense.

We shall be more accurate then if we discuss political

[12] *Ibid.*, pp. 168-69.

communication[13] as an instrument of state rather than psychological warfare, or information, or propaganda, or political warfare. Patently, a new name is needed. *Political communication*, in the long run, may not be the best name to use. But it will do to go on with, and it gets us out of our present confusion.

If *political communication* is allowed to stand as a term for all the activities carried on in the field, then psychological warfare is one branch, conducted against an enemy and against an enemy only. Information is another branch. The main element of the entire range of the instrument is propaganda, but, as it will be developed, propaganda restrained by definite premises that do not permit the unfavorable associations currently attributed to that word.

It is assumed that there is no need to argue the point that political communication undertaken by a free society must be an honorable and open activity for which government can be held responsible. Only in this context will able, competent, and honorable men be attracted to it as a career. And it is as a career that it must be considered. All the indications point to the present conflict forming the focus of international relations for a long time to come. Political communication, essentially, is a long-range instrument, and its use must be planned for and organized on such a basis.

The elements that must be considered in using this instrument should be set down so that it may be differentiated from other instruments and its field of maneuver exploited to advantage.

The point of departure is that the actions of government possess psychological consequences and therefore can be

[13] I am indebted for the term *political communication* to Dr. Leonard S. Cottrell, Jr., who first used it in an address he gave at the meetings of the American Sociological Society in Washington, D. C., September, 1955. The concept of political communication that follows is based on notes given the writer by Dr. Cottrell. While Dr. Cottrell has no responsibility for the treatment, if the concept has value, it is because of his work.

given meanings. It is the role of political communication to see to it that these actions do not go unnoticed and that the meaning given is the one intended. This, it will be remembered, is what Russia did when she launched her satellites. It is what we did *not* do when we dropped our atomic bombs on Japan. This is the main purpose of political communication. It must explain the significance of the action, and, if the action is directed to a situation outside our own borders, it must suggest possible steps that may be taken in that area that will be mutually satisfactory.

Any response by the target of our action that we may desire must be one that can be made with dignity. It follows that the way of life of the target must enter into our considerations. This means that we must understand its cultural, social, political, and psychological character. Furthermore, it is essential that we be prepared to take as well as give. Communication is a two-way process, not a monologue. We may not assume that the fountainhead of all knowledge and all truth resides inside our own borders.

Political communication must express our convictions as a nation. We know enough about the minds of men to be able to say that no trickery or smart politics is likely to persuade a people to accept and hold to a point of view coming from an alien source. The presentation of this viewpoint is much more than a matter of leaflets and broadcasts. Varieties of economic action, fleet visits, fly-overs, ways by which our citizens deployed overseas can take part in the daily lives of the people they are among, all can be forms of political communication. In fact the combinations and range of ways in which meaning can be conveyed by one government to other peoples appear to be limited only by the ingenuity of the human mind. If this is accepted, then effective political communication cannot be the responsibility of any group within a nation working independently.

The operation must be fully integrated into government and accepted as an integral part of government.

This is the scope of the instrument and, as we have employed it, it is in this direction that its history shows it has been developing. It has not yet, however, reached this point nor can it do so until agreement has been reached on the premises governing its employment and the kind of organization necessary to permit it to function to its maximum capability.

It is obvious from what has been said so far that political communication is amoral. It is equally a servant of the corrupt and of the upright. Therefore a free society may not use it indiscriminately. Its use must be governed by an ethical and political base that is supported by conviction, and this conviction must be understood and accepted by the citizens of that country. The nature of the instrument then must be set forth not in terms of factions or parties, but in terms of the national spirit and intent of a people, for on no other terms can a free people speak through its government to other peoples and expect its communication to carry weight.

We need to think hard and straight about the direction in which, as a nation, we intend to proceed. And we need to consider how we shall employ our resources to attain our objectives, for we no longer are so abundantly supplied as to be able to mount and ride off in all directions.

Until now we have muddled through. Until now we have reacted to situations as they arose. Until now we have been motivated by fear or, at best, by the stimulus of immediate challenge that leaves the initiative to our opponent.

We are faced with the task of fighting to preserve the Western heritage, not only for ourselves but for the free world. Essentially this must be a victory of the spirit, for the resort to force, on a large scale, appears destructive of the world as we know it. If we are to win this war we must capture the minds of men, and to do this we must discipline our own.

Most important, we shall not be able to do this by the use of psychological techniques. The result, if it is in our favor, will be the consequence of sustained action, action that has been planned and is expressive of a national will. It is not enough to talk of propaganda in terms of the truth of our principles. The propaganda must be the propaganda of action—definite and concrete political, material, and ethical action that may be seen and measured.

We are in conflict with an idea that "has never lost a single fundamental battle so far. In one generation it has grown to control and direct one-third of the human race. It is precisely the borderline people who are quick to draw their own conclusions from this phenomenal growth."[14] We have not yet grasped the deep implications involved in recognizing "that without profound conviction concerning the nature of things" the Western world "can never beat Communism."[15]

If, with purpose and intent, the American people

> are told they have a job to do in a generation, that the essential resources are available, that it will be the greatest achievement of man if they succeed, and that the greatest disaster that ever befell man will occur if they fail, and will also occur if they fail to try, and that there must be no illusions or nonsense about it, and if this is done from the highest point of authority as to the dangers and possibilities in the situation, they may be expected to react with all the latent idealism they have.[16]

This is not the philosophy of fear. It is positive and must be made positive by information and acts.

[14] Charles Malik, of Lebanon, in an address to government officials, Washington, D. C., September, 1954.

[15] Charles Malik, of Lebanon, in an address to government officials, Washington, D. C., December, 1952.

[16] George S. Pettee, "A Positive Position for the Third Phase of Cold War." Unpublished manuscript.

The acts must come first. It is the role of political communication to explain and interpret those acts. It calls for a militancy springing from every department of government and from the private support of individual groups within the nation that such a positive program may be expected to call forth.

In this context, political communication may expect the services of capable and dedicated men, for it will be an honorable activity. Only in this context can a free people expect it to prove fully effective.

II *The Nature of Political Communication:* THE PREMISES

The premises on which a free society may undertake political communication ought to be capable of expression, and, if they are to be operative, there must be agreement on them. So far nowhere in the writings on the subject of "psychological warfare" do these premises seem to find collected expression. Yet it is clear that they have been implicit in much of the work that has been done, and, in certain instances, they have been explicit.

Some of them differ, of course, from those on which a totalitarian state conducts propaganda. This difference probably is partly responsible for some of our confusion about the whole business of political communication. It was in the hands of Hitler and Goebbels that we saw what organized, ruthless employment of the weapon could do. In our haste to make use of it ourselves we seem, on occasion, to have equated the fact of their success with acceptance of the premises behind that success, without always pausing to reflect on whether a free society was at liberty to make this equation. The weapon is undoubtedly successful in Russian hands, but, again, a free people is required to examine the nature of this success before conducting its own operation similarly.

We can discern twelve premises that are basic to any program of political communication undertaken by a free power. They are derived from the practice of the art; from the facts of political life, both in our own society and in societies outside our borders; from the historical record of results as these are available; from the collective thinking of theoreticians and practitioners on how an art that requires scientific method in its use ought to be practiced; and from the work of social scientists in this field.

Premises derived like this cannot be considered scientifically established. But we have no science of politics. We have no theory of political science. There are many theories. The pragmatic approach is the one that we employ. It is for this reason that *premises* is selected as a better word to use than *principles*. Increasing knowledge and experience may be expected to show us refinements and alterations that, eventually, will lead to established principles. But we have not reached this point yet.

The important distinction in any attempt to state the premises must surely be the avoidance of any conflict between what our scientific knowledge tells us and what Mannheim calls the folk spirit of a people, its collective wisdom and instincts. If we can discern a set of premises that permits congruence between these two factors, we are entitled, perhaps, to consider that we have a basis on which, logically, we may undertake a national operation.

The Importance of Truth

The most significant premise to be discerned appears to be that our political communication must be based on truth. Baldly stated, like this, it invites the ribald comment that we should be against sin. Obviously truth is a somewhat nebulous concept. Simply because the premise has been baldly stated in the past, realistic operators have sometimes tended to ridicule it.

Truth is not here being summoned in any absolute sense. More nearly what is meant is that any national program in this field must be based on facts, and those facts had better be true. It is important to guard against confusing truth with psychological correctness in considering this premise. Hitler was manifestly psychologically correct in much of his propaganda. He was also demonstrably untruthful. A free society is required to be both psychologically correct and truthful, which is to say that our facts must be right.

It is obvious that any decision accepting or rejecting this premise must be in an "either . . . or" category. The importance of truth and fact must govern our political communication, or else the totalitarian principle of using the truth only when it serves our purposes must govern. In a society such as ours, composed of so many differing cultural strands, the only course that permits these strands to be woven into a source of strength for a national program in this field is the arbitrament of truth and fact. As long as it can be shown that expediency, instead, is the governing factor, there is bound to be division among the groups of opinion that make up our nation. Equally we may expect this division to be no less critical in any communication to foreign peoples. Only on the basis of truth and fact can these divisions, natural among men, be brought into harmony.

It is the nature of a democratic society to look to truth as an ally and to court it without fear. To reduce this characteristic to one of expediency for purposes of propaganda is to go against an established historical tradition. As soon as such a society learns that in its name a government agency is debasing the premise, it is reasonable to expect immediate opposition to it. This opposition will be evident to our own people and to the peoples beyond our borders, and, inevitably, the result will be a weakening of the force with which we speak.

Over and above this consideration is the position of the premise of truth as an ethic in our society. Until there is convincing evidence that we have abandoned this ethic, no agency of government conducting propaganda can hope to have much success in speaking for the nation on any less rigorous variation of it. Our record as a nation documents our regard for truth and fact, even in our failures to attain an ideal. We have no right, in any program of propaganda, to proceed on any weaker premise and expect our national credit not to suffer. If we undertake to distort, or suppress, or falsify, as a matter of deliberate, long-range policy, we are admitting our disbelief in the ability of knowledge and truth to convince men of the courses of action necessary if the human spirit is to continue its development. Not unless our national character greatly changes can we expect the American people to support a propaganda operation in its name that, even tacitly, avows this disbelief. Repeatedly our national leaders, in the context of the present struggle, have summoned this premise to our defense.[1]

If it is accepted that we are in the midst of a world revolution in thought and ideas and knowledge, only by facing facts and looking for the truth in those facts may we hope

1 ". . . the struggle in which freedom today is engaged is quite literally a total and universal struggle. . . . It is a political struggle . . . It is a scientific struggle . . . It is an intellectual struggle . . . It is a spiritual struggle . . . For this whole struggle, in the deepest sense, is waged neither for land nor food nor for power—but for the soul of man himself." Text of President Eisenhower's Address on Military and Economic Security, Minneapolis, at the annual convention of the National Junior Chamber of Commerce, *New York Times,* June 11, 1953.

Note also Chief Justice Earl Warren's call for "men and women whose minds are free, men and women who can write and speak boldly the words that set forth the truth." Address to Board of Governors of B'nai B'rith, Washington, D. C., *New York Times,* November 8, 1954.

to convince others of the validity of our position and lead such a revolution to a successful conclusion. Only on such a basis does it appear we have any right to hope we shall succeed in winning allegiance for our aims and objectives.

Lesser, but still important, support for this premise comes from pragmatic experience. Sir Robert H. Bruce Lockhart, Britain's Director General of Information in World War II, has observed that "in open propaganda inaccuracy does not pay."[2] Hans Speier has a similar warning. Propaganda "may indulge in exaggeration or minimization, as the case may be, but it must mind its limits, which lie where remote facts become immediate, that is, where stories can be checked."[3]

Propaganda based on the principle that a free people continually must search for the truth has a constructive purpose. Speier remarks that "in practicing the odious vocation of handling human attitudes, which in war gains respectability, propagandists who destroy loyalties do not get very far if they offer nothing to take their place."[4] It would appear, in the light of existing political realities, that as a long-range policy any attempt to influence men's minds, lastingly, can be only in terms of determinable facts.

What weapon is there with which to combat the ideology that confronts us but the sober determination of facts? On what principle can a free people attempt to thwart the Soviet conception of the slave state except on the basis of truth, offering deeper loyalties in return for those destroyed? And, furthermore, with such a principle, what Speier calls the "odious vocation" of the propagandist loses its odium. It is important that this taint attaching to propaganda be re-

[2] Sir Robert H. Bruce Lockhart, K.C.M.G., "Political Warfare," *Journal of the Royal United Service Institution* (London, Whitehall), May, 1950, p. 197.

[3] Hans Speier, "Morale and Propaganda," in Daniel Lerner (ed.), *Propaganda in War and Crisis* (New York: George W. Stewart, Inc., 1951), Ch. I, p. 14.

[4] *Ibid.*, p. 18.

moved. As Mannheim so clearly shows, if it is merely a question of conflicting ideologies, the taint remains. It is only the objective search for truth that can remove it.

Given the moral and political structure of our society, the logic of the premise of truth seems inescapable as a foundation for any long-range program of political communication. Although by no means perfectly executed, the history, the literature, and the practice show that it has occupied a cardinal place in the propaganda operations conducted by a free people. It may be noted that a free people may not forget, either, that the very nature of its society requires this premise since it is "The habit and necessity of free people . . . to question, criticize, object and demand information about every move and where it leads."[5] This is the sort of climate in which facts will out, and, under these circumstances, any attempt on the part of government to obscure facts to a target possesses the inherent possibility of recoiling by virtue of the actions and utterances of free citizens.

The premise that our political communication must be based on truth rests on solid ground. This premise is not concerned with any concept of truth in a narrow, absolute sense. The referent is the combined moral and political principles of our society as these have been demonstrated by our history and heritage. In this sense we are required to undertake political communication in terms of facts and to give evidence by our actions that we are receptive to facts.

Under this premise we are not stripped of our ability to maneuver by a cliché of truth. The feint in a fight is a legitimate exercise of discretion and judgment and will be so accepted and recognized by opinion. But we are prevented from resort to the lie in an argument. It is the integrity of this position that is buttressed by the premise, and it is this

[5] Anne O'Hare McCormick, "The Dual Role of American Diplomacy," *New York Times*, February 28, 1953.

integrity, maintained, that will persuade public opinion of the maturity and stability of our free institutions and values and prove that reliance may be placed on our leadership.

The Force of Ideas

The second premise emerges with equal clarity and distinctness. It may be called a conviction of the importance of ideas. This premise rests on the verdict of authority, history, and experience. It is clear, in any serious study of the subject of propaganda, that ideas occupy a prominent place. The evidence also shows that if these ideas are to sway and dominate the minds of men they must be much more than mere intellectual exercises. They must be capable of being grasped by the untutored and ignorant, although intelligent. In fact, it has not been possible to think of any of the great and powerful ideas of the world that do not possess this characteristic.[6]

If we look at the record of our actions since the end of World War II, we will find in it much to support the validity of Lasswell's observation that "We are in a war of ideas but we have not found our ideas."[7] Charles Malik of Lebanon

[6] In the hands of Hitler this premise was clearly accepted. "The Nazis do not attach any intrinsic value and any stable meaning to the ideas which according to their claims are important to them. The elasticity of those ideas which are most intimately connected with Nazi ideology proves that these ideas, too, are used instrumentally. These ideas, too, are subordinate to what really matters to the Nazis: success and power." Hans Speier and Margaret Otis, "German Radio Propaganda to France during the Battle of France," in Daniel Lerner (ed.), *Propaganda in War and Crisis,* Ch. 10, pp. 239-40. The point to notice here is the controlling force of the first premise discussed. In a totalitarian state that premise is expediency. In a free society it is truth. The ideas of a free people, therefore, may not be subordinated to ends in conflict with this premise without loss of conviction. But, in either instance, the force of ideas is apparent.

[7] Harold D. Lasswell, "Policy and the Intelligence Function," in Daniel Lerner (ed.), *Propaganda in War and Crisis,* Ch. 4, p. 67.

has expressed this same conviction, but in greater detail.[8]

Demonstrably we are not generating the ideas that are needed to capture the minds of men. Yet it is ideas that are running like flame in dry grass across the world. The ideas of communism and nationalism dominate large numbers of the earth's population. And we are in conflict with them. We have met them with guns and with gifts of "butter." They have not lost their force. And yet "The field is left entirely open to us. We have an absolute monopoly in it. No Communist government, no Fascist government, no government that is totalitarian could tolerate" an approach based on "Man's right to knowledge and the free use thereof." The totalitarian could not "tolerate institutions in which such a spirit abided, because the teachers and students, uninhibited in their pursuit of ideas, would soon generate ideas that would cause the totalitarian state to crumble."[9]

We may not forget the observation of the Senate committee that investigated our information services:

> Too much material is devoted to negative anti-Communist attack. . . . people want "the simplest explanations of what democracy is and how it works" and a "positive demonstration of the superiority of democratic life and ideology."[10]

[8] "I must say in all humility that the leadership of the West in general does not seem to be adequate to the unprecedented challenges of the age. There is a tragic dearth of men, men who are so genuinely in touch with the truth and with the hearts of their fellow men as to have only to open their mouths to be loved and believed and followed. . . . There is a corresponding bankruptcy of fundamental ideas." Malik in "War and Peace," a statement made before the Political Committee of the General Assembly of the United Nations, November 23, 1949, The National Committee for Free Europe, Inc., New York, p. 30.

[9] Chief Justice Earl Warren, "Free Investigation and Faithful Experiment," address prepared for delivery at the dinner of the Association of the Alumni of Columbia College, January 14, 1954, *New York Times,* January 15, 1954.

[10] Report of the Committee on Foreign Relations, Overseas Information Programs of the United States (Hickenlooper Report), Senate Report No. 406, Eighty-third Congress, First Session (1953), p. 175.

Political Communication and National Policy Must Be in Accord

It is not nearly enough for propaganda to be true to the moral and political principles of the nation and to be able to produce political conviction. It must, in addition, be congruent with national policy and national objectives. As far as propaganda beyond our own borders is concerned this means congruence with foreign policy, and "A foreign policy, by its very nature, needs to be a continuing policy. And if it is to continue, it has to be a national policy, not just a party policy."[11]

The importance of linking propaganda to policy is now widely recognized although in World War II this was by no means always the case. Since then, however, there have been a number of studies, papers, and articles emphasizing its necessity. In essence their substance may be briefly summed. Although there may be distinct psychological plans and specific psychological activities directed toward national objectives, there are no national *psychological* objectives separate and distinct from national objectives. Overt propaganda conducted by a free society is necessarily based upon policy and has no life apart from it. The psychological aspect of policy is not separable from policy, but is inherent in every diplomatic, economic, or military action. There is a psychological implication in every act, but this does not have life apart from the act.

This reference to the psychological aspects of diplomatic, economic, and military action directly links such action to the concept of political communication that is far wider in scope than the term *psychological warfare* allows. And, in practice, psychological warfare has clearly refused to be

[11] Text of address by former President Harry S. Truman before the Reserve Officers Association, Philadelphia, *New York Times*, June 27, 1953.

bound by the restrictive connotation of the term in general use. Diplomatic and economic action is constantly oriented toward allies and neutrals. We cannot properly undertake "warfare" against them, but political communication as a concomitant of such action is both permitted and desirable.

National policy self-evidently sets limits to our propaganda, and just as evidently any propaganda that departs from national policy will be contradicted by our actions and its effectiveness nullified. Within the limits of national policy, propaganda must be based on the truth and speak with political conviction. It obviously cannot go beyond these limits in either direction. Equally, it cannot operate in a frame of reference that omits them.

It is the policies and acts of government that furnish psychological warfare with its ammunition. Of significance, too, is the wording and time of release. These are political in nature and beyond direct military control, but those who make or influence policy need to understand the nexus between political, diplomatic, economic, and military decisions and the success of any psychological warfare objective.

It is on this base that the main argument for civilian control of political communication rests. "The exclusive concern of the military bureaucrat is military action and, if that proceeds according to plan, then all the rest of life is in order too."[12] Political communication is concerned with far more than military action.

> Who, precisely, in the opposing nation is the enemy? . . . Is it the foreign policy of the other nation that is to be modified or also its social institutions? Precisely which, if any, of the latter; and in what way? . . . In war, what are the political long-range objectives, if victory is a medium-range aim? . . . It is answers to questions of this kind which furnish the basis of political warfare, as dis-

12 Karl Mannheim, *Ideology and Utopia* (London: Routledge and Kegan Paul, Ltd., first publication in England, 1936), p. 106.

tinguished from the gossip of news, the preaching of ideology, the performance of tricks, and the projection of the self.[13]

In a democratic society the answers to such questions must be given, if they are given at all, by political, as distinct from military, authority even though military considerations may be a governing factor.

The manner in which national policy may limit propaganda is perceived when two examples from the two world wars are compared. Contrast the appeal of President Wilson's Fourteen Points in World War I with the negative effect of the slogan "Unconditional Surrender" in World War II.

The argument may be put in the following words: A planned campaign of political communication is not planned unless policy and objectives have been determined, for without policy and objectives who can speak with political conviction? George Taylor, who had considerable experience with this instrument in World War II, has summed up our developing recognition of the interaction between our policy and our propaganda:

> The problem today . . . is not so much one of securing acceptance for the use of psychological warfare, but rather one of securing an understanding of what it means and can do. . . . It cannot be stated too strongly that psychological warfare is an arm, not only of military operations, but, like military operations themselves, of national policy. The national policy is not exclusively a civilian matter in the realm of application, but it is most definitely a civilian affair in formulation. It is the business of the elected representatives of our country. The acceptance of this view would not, however, solve the problem. It does not really matter very much whether psychological warfare is conducted by men in uniform or out of uniform so long as the basic ideas that the oper-

[13] Hans Speier, "Psychological Warfare Reconsidered," in Daniel Lerner (ed.), *Propaganda in War and Crisis,* Ch. 27, p. 483.

ators hold are the same. These ideas must include the concept of psychological warfare as implementing over-all national policy. . . . During the Second World War the theater commander controlled practically everything that went on in his theater of operations. In effect, therefore, theater commanders made political decisions. But who can say that theaters of operations in the next conflict, if it should come, would be the same as those in the last? . . . So we have to look at this old problem afresh and seek its solution not only on the jurisdictional level but rather in the development of a view of national policy common to both those who make and those who implement policy. Psychological warfare divorced from national policy is ultimately as ineffective as it is dangerous.[14]

The Factor of Public Opinion

The fourth premise springs from the controlling influence of public opinion. This premise does not apply, of course, to propaganda conducted by a totalitarian society. It is a powerful factor in propaganda conducted by the government of a free people. On reflection this must be so if it is accepted that governments derive "their just powers from the consent of the governed." It is a position commonly held by Americans who have had experience with this activity that in the last analysis they have been bound by the state of public opinion in their country. The premise, however, does not weaken initiative.

It is a truism that public opinion is not immutable. It is equally true, however, that it is not possible to accomplish a plan requiring its support if the leadership is too far in front of, or behind, popular understanding. Nevertheless the force and impact of public opinion, even when latent and unaroused, condition government action in the field of po-

14 George E. Taylor, "Rearming with Ideas," *Virginia Quarterly Review* (Charlottesville, Virginia, University of Virginia), Spring, 1951, pp. 162-64.

litical communication. The issues on which public opinion will assert itself, or, more properly the directions it will take in asserting itself on issues, are not predictable. No one will attempt to deny that when it chooses to make this assertion, it is decisive.[15]

Public opinion is like an amorphous substance; it is in a state of continuous existence, but without leadership it does not crystallize. On the occasions when it does crystallize, statesmen, presidents, prime ministers, and generals rise either to still greater heights of leadership or run for cover. The man who stands against this force is a very brave man indeed, and when on political or moral issues such a man is in evidence, if he wins, it is by reason of "the spirit of a people, and the strength and order arising therefrom," and not solely because of personal power.

The propagandist, therefore, may not conduct his operation without at least the latent support of this public opinion. But, in addition, he must have some apperception of the spirit of his people in order that he may lead. And, in leading, he will be aware that he must challenge habits of opinion—one of the hardest and most intractable of all habits to modify or change. At the same time he knows that he cannot lose touch with the "folk spirit" and that, in a free society, this folk spirit is, in fact, the hand on the reins.

[15] "The inner form of historical individuality existing at any given time, be it that of an individual personality or of a folk spirit, and the external conditions together with the past that lies behind it, determine the shape of things that are to be." To which Mannheim adds a footnote by Adam Muller: "There is no substitute for the spirit of a people, and the strength and order arising therefrom, and it is not to be found even in the brightest minds or in the greatest geniuses." Karl Mannheim, *op. cit.*, p. 210.

Political Communication Is Rooted in Action

The premise that action is essential to effective political communication is one that has long been well established. In practice it has been, on occasion, disregarded. The relation between action and political communication is direct and powerful. Both the history and all our experience with the instrument conclusively show that, if it is to be effective, what a people says must be backed up by what a people does. In this field, pre-eminently, it is no good to make well-meant gestures in the direction of good will unless they are implemented. It is useless to pay lip service to ideals unless it can be seen—and seen by the onlookers—that the ideals are a motivating force governing what is actually done.[16]

Masters of political communication have always understood this, and in addition to being able to fire the hearts of men with words they have been capable of performing the deeds that gave those words force. It can never be forgotten that what the propagandist is after is action. He is

[16] Lynd succinctly points up the significance of this relationship when he says, "Democracy as symbol and slogan is weak unless democracy as operating reality is strong." Robert S. Lynd, *Knowledge for What?* (Princeton, N. J.: Princeton University Press, 1948), p. 236.

Lerner makes the same point: ". . . the most effective action is that which integrates words and deeds to a common purpose." Daniel Lerner, "Effective Propaganda: Conditions and Evaluation," in Daniel Lerner (ed.), *Propaganda in War and Crisis,* Ch. 21, p. 346.

Barrett, who has had to think about these problems from behind the administrative desk, says, ". . . the best propaganda is action intelligently executed and then fully publicized through information channels." Edward W. Barrett, *Truth Is Our Weapon* (New York: Funk & Wagnalls Co., 1953), p. 206.

Finally, there is that master of the art, Sir Winston Churchill, of whom Forrestal records: "He said that a gesture of power not fully implemented was almost less effective than no gesture at all." Walter Millis (ed.), *The Forrestal Diaries* (New York: The Viking Press, Inc., October, 1951), pp. 144-45.

not content merely with holding the revival meeting; he wants to see the sinner hitting the sawdust trail.[17]

Although the weapon of the propagandist is the word, it is by no means the word alone. It is the word as it explains and interprets and prophesies action. It is this very premise of action that may well, in the long run, turn out to be the pivot on which the defeat of communism will turn if it is accepted that the Communist system will not be able to live up to its promises. In other words, its propaganda will not be supported, over time, by its action. Implicit here is the requirement that in opposing communism we deliver something better.

Goebbels, as we might expect, saw the interconnection between words and action realistically and set it down in his diary: ". . . at the moment we cannot change very much through propaganda; we must once again gain a big victory somewhere."[18] Against what must have been his better judgment, he attempted the impossible, only to fail, and in failing, underscore the importance of the premise. "After the sweep of the allied armies across France in 1944, for example, it was extremely difficult for Goebbels to convince

17 ". . . the true propagandist is almost invariably a man of action. His words are directed toward action, are indeed a form of action, and he constantly tends to participate in the action which his words inspire." Gorham Munson, *Twelve Decisive Battles of the Mind: The Story of Propaganda During the Christian Era, with Abridged Versions of Texts That Have Shaped History* (New York: Greystone Press, 1942), pp. 33-34.

"The objective of propaganda is action, not merely readiness to respond. External rather than internal public opinion is sought. The learned attitude—the pre-action response—must affect behavior. It must lead to what may be called an action response. The advertiser . . . induces [people] to buy. . . . The politician seeks more than admiration . . . actual votes count. The war propagandist . . . wishes the enemy to feel the hopelessness of their cause so that they will fight or produce less vigorously and eventually surrender." Leonard W. Doob, *Public Opinion and Propaganda* (New York: Henry Holt & Co., Inc., 1948), p. 397.

18 Leonard W. Doob, "Goebbels' Principles of Propaganda," *The Public Opinion Quarterly* (Princeton, N. J.: Princeton University Press), Fall, 1950, p. 442.

the German people that 'we shall win but we can lose. . . .' "[19]

Events, "things happening," action—each and all of these words may be used to exemplify what is meant by the importance of action to political communication. The right acts combined with the right words possess tremendous leverage. Take the right action, but give the wrong reason, and there is trouble. The importance to propaganda of action is a cardinal premise, to depart from which is to invite a weakening of effort. The propaganda of the deed stands out in any study of the premises of political communication as a firm foundation underpinning any national program.

Political Communication Cannot Be More than an Auxiliary Instrument

There is a corollary to the premise that political communication is rooted in action, which in the past operators have at times ignored. In fact, during World War II, operators argued in denial of it. The corollary is: Political communication is never more than an auxiliary weapon. It is important not to lose sight of this factor. In the past, attempts have been made to build the instrument up as a primary weapon, as responsible for this number of surrenders or for the capturing of that stronghold, for this action by an enemy people or for that change of attitude on the part of neutrals. Partly, this has been so because of the connotation the activity carried of being something dishonorable except in time of war, when many things dishonorable in peace become acceptable, and even laudable, practices. Honorable men naturally want to feel that the work they are undertaking is legitimately important and essential. Because of the demonstrable hostility that has existed toward propaganda, because of the demonstrable attempts

[19] Leonard W. Doob, *Public Opinion and Propaganda,* p. 36.

that have been made to belittle it by men in more established professions, there has been an understandable attempt on the part of men working in political communication to build up the ego of this modern accessory to the arsenal of weapons.

There is no longer any need for apologetics in discussing political communication. It works. And there is sufficient evidence of this, sufficient history, sufficient knowledge, for the student and practitioner not to need to worry in the presence of skepticism. It is a firm instrument of national policy in time of war and cold war. What is important is to understand and learn more about its most effective employment. Facts are necessary here. There is no longer any need to argue for its acceptance. It has won its place. The task is to see that it occupies this place to its utmost efficiency. It cannot do this if it is thought of as a primary factor. Its function is as an auxiliary to the main weapons, a vital auxiliary nevertheless, capable, as the Drogheda Committee[20] said, of being decisive in tipping the balance.

The Indispensability of Knowledge

A further premise that emerges from a study of political communication is concerned with knowledge. It is, however, a specialized kind of knowledge that may be divided into two sorts. As Lockhart observes: "Political warfare is neither an exact science nor a separate art."[21] Nevertheless, scientific knowledge is indispensable to successful practice. This is, in itself, self-evident. It is the kind of knowledge that is important.

20 Summary of the Report of the Independent Committee of Enquiry into the Overseas Information Services (appointed by the British Government, 1952, under the chairmanship of the Earl of Drogheda, P.C., K.C.M.G.), April, 1954, Miscellaneous No. 12 (1954), Her Majesty's Stationery Office, London.

21 Sir Robert H. Bruce Lockhart, K.C.M.G., "Political Warfare," *Journal of the Royal United Service Institution* (London, Whitehall), May, 1950, p. 195.

> If nobody goes about endlessly counting throughout a lifetime the number of particles of sand along infinite miles of seashore over all coasts of the world, why is this? Because there is no point to it, no need to complete this particular aspect of the jigsaw puzzle of the unknown.[22]

The scientific knowledge required is the kind provided by the social sciences. It consists of a thorough comprehension of the sociology of given targets—their cultural, political, economic, social, and psychological patterns.

The operator must be aware of the climate in which he has to do his work. He cannot expect to do an adequate job if he does not possess knowledge of the political, social, intellectual, and ethical trends of significance and importance in the time span that concerns him. This does not mean only the trends in the target society; there is an equal requirement for being able to interpret the trends as they manifest themselves in his own society. Linebarger has held that propaganda is most effective when it is in phase with the forces affecting the climate of mood and opinion in both the "sending" and "receiving" societies.[23]

It is much easier, of course, to set this down on paper than it is to provide for its integration into a planned operation. Men who can single out and interpret trends in these spheres are not common. The skill is not one that depends solely on learning or discipline. It is, in part, a faculty born with an individual. It requires sensitivity. Training and development of the faculty increase its scope, but there is no course in any university that guarantees that those who take it will graduate with the ability to discern and interpret the climate of international situations as they arise, and thus increase the prospect that the right words will be used, words that will induce the requisite action.

22 Robert S. Lynd, *op. cit.*, p. 183.

23 Paul M. A. Linebarger, author of *Psychological Warfare* (Washington, D. C.: The Infantry Journal Press, 1948), and a leading writer and consultant in this field.

A classic example is provided by Germany's psychological warfare against the Russians in World War II. The policy of Hitler did not permit the German army to exploit the psychological situation that existed. In default of an operation that was in phase with both the Russian and the German moods, Germany was unable to take advantage of the climate her soldiers found when they occupied Soviet soil.

This failure to interpret, and act on the interpretation of, the climate is not, of course, the sole or even the main reason for the failure of German psychological warfare. It was a contributing factor, however. And a study of the Wehrmacht files shows that there were those in the German army who recognized this.

It is to the social sciences that the propagandist will need to turn for help in discovering and interpreting the changing world climate, although it is by no means to the social sciences only.

It should be a basic proposition of political communication that the target and the goal must be defined and specified. Probably no one with any experience will disagree with this as a premise. It is, unfortunately, equally probable that with the statement of the premise all further agreement will cease. When it comes to implementing the proposition, opinions are a dime a dozen.[24]

There is considerable support for the view that propa-

[24] Lockhart says propaganda "should be addressed to the masses." Sir Robert H. Bruce Lockhart, *Comes the Reckoning* (London: Putnam and Co., 1949), p. 154.

Speier has written: "Since in modern societies the mass of the population cannot overthrow, or actively influence the policies of, despotic regimes without armed domestic or foreign support and without organized leadership, the population at large is no rewarding target of conversion propaganda from abroad. Any notion to the contrary may be called the democratic fallacy of democratic propagandists who disregard the differences in political structure between the regimes under which they and their audiences live." Hans Speier, "Psychological Warfare Reconsidered," in Daniel Lerner (ed.), *Propaganda in War and Crisis,* Ch. 27, p. 473.

ganda is something for the masses, not for the elite. In World War II there was a good deal of muddy thinking and action on this point. In the Korean war the target and the goal do not seem to have been clearly set forth. The Operations Research Office found that the goal, on the evidence of the type of propaganda disseminated, was surrender. But surrender of which? The Chinese officers? Enlisted men? North Korean officers? Enlisted men? Not until certain studies undertaken by the Operations Research Office had been completed does there seem to have been any clear evidence of the best target to go after.

In logic it seems that we shall get nowhere in looking for a course of action until it is first decided whether we want the target "converted" or whether we want what Speier calls "relevant political action." The choice is between these two alternatives. On the face of it conversion offers far less as a springboard than does political action. For one thing there are elements of mysticism and religion in the notion of conversion, and to these must be added a tincture of fanaticism. We have to ask ourselves if our present climate would support attempts at conversion.[25]

[25] Lynd says: "Religion, in its traditional forms, is a dying reality in current living. And yet no culture can live without a central core of emotionally resonant loyalties widely shared by the mass of the people. . . . American culture, if it is to be creative in the personalities of those who live it, needs to discover and to build prominently into its structure a core of richly evocative common purposes which have meaning in terms of the deep personality needs of the great mass of people." Robert S. Lynd, *op. cit.*, pp. 238, 239.

Nock, in discussing the growth of religious movements, says: "All these things we see as movements governed and directed by political and other considerations conditioned by the intellectual atmosphere of the times. All of them turn largely on difficult and nicely balanced decisions made by individuals on the questions of man's relations with the unseen and the unknowable. All of them we know in part and understand in part. Which of us has any accurate idea of what is going on in the mind of the average churchgoer. How far does he himself know? And if this is so with the

The choice of conversion opens a door into a realm of the human spirit marked by divergencies of thought and practice. In the final analysis, as Nock says, the issue comes down to an opinion about the ends of man.[26] If the choice of goal is conversion, we are bound to take sides, and to the extent that public opinion is divided our propaganda will suffer.

It seems impossible to controvert that in the mid-twentieth century we are in the midst of a tremendous conflict in the human mind. The ancient authority of the priesthood has

present, how much more so is it with the past. . . . One of the greatest of scholars has said: 'The tradition yields us only ruins . . . the tradition is dead; our task is to revivify life that has passed away. We know that ghosts cannot speak until they have drunk blood, and the spirits which we evoke demand the blood of our hearts. We give it to them gladly; but if they then abide our question, something from us had entered into them.' Anyone who has tried to get at the realities of ancient thought and belief knows how true this is. Our interpretation must be in a considerable measure a personal interpretation, based on a balance of probabilities but informed by the spirit which we bring to it." Arthur D. Nock, *Conversion: The Old and the New in Religion from Alexander the Great to Augustine of Hippo* (Oxford: The Clarendon Press, 1933), pp. 269-71.

[26] ". . . the doctrine of grace. The genius of Christianity lies on the side of Augustine, the genius of paganism on the side of Pelagius. The one built on a consciousness of sin and on revelation, the other on a consciousness of goodness and on common sense. On this issue we must all take sides." *Ibid.*, p. 249.

In this connection Speier and Davison have done some pertinent thinking. They have reminded us of those who hoped at one time it would be possible to convert the world to democracy, through superior communication techniques. Slogans were devised calling for a "Truth Bomb," a "Campaign of Truth," or a "Crusade for Freedom." These slogans stemmed from historical images, for example, the work of Christian missionaries in converting the heathen, the Moslem armies who fought for victory and paradise to fulfill the promise of the Koran, or the surge of liberalism and democracy in the eighteenth and nineteenth centuries that swayed the civilized world. ". . . these dynamic ideas of the past, however, [were not] devised by a national government as an instrument of policy." They sprang "from the minds of visionaries and philosophers." It was not until their power was apparent that "they were *adopted* as a means of policy by one or more states. Dynamic ideas are difficult to think up on order." Hans Speier and W. Phillips Davison, *Psychological Aspects of Foreign Policy* (Santa Monica, California: The RAND Corporation, 1954). Out of print.

been challenged by the facts of science, and scientific method has, to some extent at least, supplanted fiat. Where men will turn for the final answers to the questions and problems they must deal with varies among individuals, among groups, and among nations. To attempt anything like conversion in such circumstances seems foolhardy. Opinion on the wisdom of any course elected would undoubtedly be divided. And, particularly in our own case, we may forget Karen Horney's observations on the conflicting American characteristics of brotherly love and aggressive competition.[27]

If the propagandist is in favor of brotherly love, no one will take him very seriously, and our actions undoubtedly will not always square with his precepts. If he favors aggression, he will be in trouble. The intelligent compromise seems the only sensible course. This must bar conversion in any form. It does permit agreeable political action. This is not to say that our aims have changed. It is to say that in a pragmatic world working solutions will need to be accepted until, in a free society, once again there is sufficient unanimity of opinion to permit greater missionary zeal for conversion. Venturing a prophecy, this zeal does not seem very much in evidence, and to count on its reappearance in the

27 ". . . the question arises as to whether there are not likewise certain definite contradictions in our culture, which underlie the typical neurotic conflicts. . . . The first contradiction to be mentioned is that between competition and success on the one hand, and brotherly love and humility on the other. On the one hand everything is done to spur us toward success, which means that we must be not only assertive but aggressive, able to push others out of the way. On the other hand we are deeply imbued with Christian ideals which declare that it is selfish to want anything for ourselves, that we should be humble, turn the other cheek, be yielding. For this contradiction there are only two solutions within the normal range: to take one of these strivings seriously and discard the other; or to take both seriously with the result that the individual is seriously inhibited in both directions." Karen Horney, M.D., *The Neurotic Personality of Our Time* (New York: W. W. Norton & Co., Inc., 1937), p. 288.

immediate period facing us appears unrealistic. Political action that is relevant appears as a goal more acceptable for majority support, and political communication by a free power requires majority support.

It is not clear from a study of political communication whether this first and most important decision has yet been made by the political authorities of the United States. It is not clear, even, that the necessity for making it has been faced.

What is the goal of our propaganda to the world? Some of our activities appear to lend support to the hypothesis that we are not seeking only political decisions that are congenial, but actual conversion to our own way of life. If there is, in fact, this dichotomy, we must surely be lessening to some degree our ability to gain one end or the other.

What is our aim? Our national objectives are to be found in the appropriate documents. Is our propaganda to try to convert other peoples to our own conceptions or is it to seek political action by them that is congruent with our own beliefs and policy? Political action that is relevant appears as a far more realistic approach, but existing papers seem equally to permit an attempt to convert, and whether the goal is conversion or relevant action probably depends, not on a policy decision, but on the personal bent of the operational staffs putting the messages together. The policymakers do not appear to have formed a definite conclusion.

Once the goal is clear, the target calls for definition. Whom do we want our messages to reach, and why? Is what we are asking feasible? Does it endanger our friends, if any, in the target? Political communication cannot expect results if it is haphazard. Its activities themselves establish the premise: The target and the goal must be defined and specified. For this, our only reliable resource is knowledge.

Experience has shown that the internal and external forces existing in the target condition the operation to be under-

taken. The validity of this is evident on reflection. The form of the message selected for a people with a low standard of living is unlikely to be the same as that for a people technologically advanced. The message itself may be the same, but the manner in which it is presented may be expected to vary. In fact, if the premises under discussion are borne in mind, there is a requirement for the messages to be related to each other. The requisite political action desired in either instance ought not to contravene the truth of the political and moral principles activating the originating society.

The importance of this premise is recognized by the enemy. Milosz gives a vivid picture.

> In every capital of Central and Western Europe the windows in the Central committee buildings are illuminated late into the night. Behind their desks sit men well versed in the writings of Lenin and Stalin. Not the least of their tasks is to define the position of the enemy. . . . Data from each country then aid the supreme command in Moscow to establish an over-all strategy. Different groups of people are the main object of study.[28]

The situation in the target is bound to condition decisions taken at the firing line. Knowledge of this situation is one of the prerequisites to deciding the specific content of messages.

Political communication now has what may quite properly be termed a body of literature in the proper scientific sense. The social sciences have clearly established their ability to provide the operator with tools and concepts that give him greater assurance of how to proceed. The sort of knowledge required, in a field so largely filled with "unknowns," requires, among other things, skill in differentiation. ". . . the outstanding characteristic of a well-trained scientist is

28 Czeslaw Milosz, *The Captive Mind* (New York: Alfred Knopf, Inc., 1953), pp. 191-92.

his ability to distinguish 'significant' from 'insignificant' problems and data."[29] And most particularly in this field, probably, there exists a requirement for imaginative synthesis.

> The scientific temperament feels much more comfortable when it is breaking down a complex phenomenon into simpler parts than when it is trying to pull together a series of diverse facts into a unity of relationship. For a solution of the ultimate riddles, however, synthesis is more important than analysis. . . . It is not an understanding of units which we now seek, but of unity.[30]

It is not knowledge, in the sense of intelligence, that is required in the context of this discussion. It is knowledge in the scientific context: knowledge of what stimuli can be counted on to accomplish desired action responses; knowledge of what cultural factors can be changed in what time space; knowledge of what cultural factors the government of a free people ought not to try to change; knowledge of the drives and motivating influences that are most important in a culture alien to our own.

Knowledge of the technical accomplishments by which ideas can be transmitted from one place to another we largely possess. But knowledge of the effect these ideas may have when introduced to the target is far more limited. Under the concept of political communication, we particularly require cultural knowledge and a refinement of the tools and techniques by which such knowledge is acquired and tested for accuracy. Only as we approach greater knowledge of how the mind works, of the importance of folk ways, of the inertia attendant on the introduction of new ideas, of the drag of tradition, shall we be able to concentrate instead of diffusing the impact of messages we may desire to send.

29 Robert S. Lynd, *op. cit.*, p. 183.
30 *Ibid.*, p. 245.

So far, in discussing the prime importance of knowledge, the emphasis has been on the contribution to be made by the social sciences. A second sort of knowledge was also postulated. This consists of information about our own plans and policies. The first sort of knowledge is difficult to acquire. In many respects such cultural knowledge as we know how to obtain is imperfect because of insufficient data and because we have not yet discovered the right tools for acquiring it. The second sort requires only a command decision. Yet without it we are not able to engage in effective political communication. The whole history and practice of the activity document this statement. The propagandist in the past has not been told what our own plans and policies were. The policy of "unconditional surrender" in World War II, for example, came as a complete surprise to our propagandists. At the time the decision was made to drop the first atomic bombs on Hiroshima and Nagasaki no attempt was made to try to find out the likely psychological consequences, and, again, our propagandists were uninformed until after the event. Still later, our whole procedure with respect to the development of nuclear weapons and their resultant inevitable impact on policy has been kept from those primarily engaged in political communication. The situation is better as of this writing, but it is still far from satisfactory, and it is still inhibiting our political communication at a critical period in our history. Without knowledge of our own plans and policies, however precise our sociological intelligence becomes, we cannot expect successful political communication.

In general we are better equipped now than at the close of World War II. There at last exists a transmissible fund of knowledge to which practitioners may turn. Much has been collected and ordered. Never again should it be necessary, as it was in the two world wars and the Korean conflict, to undertake psychological warfare without accepted premises on which to plan, make decisions, and organize. Nevertheless we are still only on the threshold. It is not knowledge about psychological warfare only that we need. The concept of political communication requires an increase in our basic sociological knowledge, and this is an area as yet unorganized so that we do not even possess the knowledge enabling us to determine priorities of research.

In a struggle in which intense research activity centers on physical weapons, the case seems cogent for similar intensive research in the use of an instrument designed to influence men's minds without corroding or undermining the loyalties, principles, and values of a society whose foundations rest on a deeply held conviction that the individual is the ultimate unit of value. Only if knowledge is used constructively in support of this ideal can a free society expect honorable men to make a profession of political communication. And only in this sense can we expect it to achieve lasting results. It is knowledge in the sense of truth and fact, as facts may come to light and alter existing concepts in the light of scientific advances, to which we must look if we would win the contest for men's minds. Knowledge used as poison, to achieve temporary advantage, to corrupt and arouse hatred, merely puts one ideology in conflict with another ideology with a resort to force as the logical outcome.

Action by the Target Is the Test

If a program of political communication is undertaken, it is, obviously, in the hope of getting certain results. These results must be in the form of action responses. Attitude changes precede action responses and are, therefore, significant as the forerunners of action. Nevertheless, a planned propaganda operation obviously cannot be content with merely an attitude change. It is the overt action that establishes the fact that an attitude has changed, and, until this action is forthcoming, the propagandist cannot know whether he has succeeded or not. Since the operator is concerned, essentially, with words, it would seem well in this connection to keep in mind Dewey's observation that "there cannot be any assurance that any given word when used by two different persons even refers to the same thing, since the thing is not open to common observation and description."[31] The proof in propaganda comes when something takes place that is open to "common observation." If this action squares with what was intended by the words used, we may assume that the terms of reference are close enough to justify considering that meaning has been conveyed. But until there is this open and verifiable result, we do not know. We can only guess. The practitioner, it seems clear, is forced to take the premise that he is seeking action responses, and only when these are in evidence may he conclude that his job has been done.

The Problem of Talking to One Another

Knowledge of the forces existing in the target automatically requires us to face the problem of talking to it if there is to be any effective communication. This is not a matter of facilities. It means the ability to speak, or write, and be understood at the other end. Mere knowledge of the language does not ensure this communication. Moreover it is not only peculiarities in the target that may block understanding. It may well be that the ideas held about the people to whom we wish to talk are such that we are prevented from express-

31 John Dewey, "Theory of Valuation," *International Encyclopedia of Unified Science* (University of Chicago Press, 1939), Vol. II, No. 4, p. 10.

ing ourselves in terms that allow a mutually sympathetic understanding.[32]

It is clear from what has been written on this subject of communication that considerable thought has to be given to ways and means of establishing it. Unless two or more people can understand each other, mere words cannot be expected to accomplish much. Pettee has said that "We find situations on occasion where plenty of communication occurs, as measured by words or data, and yet where it is plain that neither the sender nor the receiver understands the situation."[33] This leads easily to what Wirth calls men talking past each other.[34]

The operator cannot afford to talk past his target. The essence of his activity is understanding and comprehension, and his aim is to shake the thought processes of the people to whom he speaks. "Horizontal mobility," Mannheim says, ". . . shows us that different peoples think differently. . . . Only when horizontal mobility is accompanied by intensive vertical mobility . . . is the belief in the general and eternal validity of one's own thought forms shaken."[35] Mannheim is talking here of movement, not only from country to country, but from stratum to stratum within countries. Political communication needs to act with this awareness constantly in mind. It deals with peoples whose vertical mobility may

[32] Barrett calls it an old lesson "that if people think in terms of villages, it is necessary to talk to them, not in cosmic terms but on a village level." Edward W. Barrett, *op. cit.*, p. 50.

[33] George S. Pettee, The Walgreen Lectures, "The Soldiers and the Republic," Lecture VI: "Responsibility." University of Chicago, January, 1954. Unpublished manuscript.

[34] "The absence of a common apperception mass vitiates the possibility of appealing to the same criteria of relevance and truth, and since the world is held together to a large extent by words, when these words have ceased to mean the same thing to those who use them, it follows that men will of necessity misunderstand and talk past one another." Karl Mannheim, *op. cit.*, Preface, p. xxvi.

[35] *Ibid.*, p. 6.

not have been equal to our own, although the rapid improvement in the technological side of communication means that in the world today we have experienced, in the past fifty years, far more horizontal mobility than in all the period of prior recorded history.

One of the results, as Barrett points out, is that "foreign relations are no longer relations between personal sovereigns or sovereign governments, but are more than ever before relations between peoples. Governments are being increasingly influenced in their conduct of foreign relations by the peoples of the world."[36] Effective political communication, then, must be able to appeal in a language that will be understood.

Milosz indirectly points up this problem of communication in showing something of what it was like to live in defeated Poland.

> Still we lived; and since we were writers, we tried to write. True, from time to time one of us dropped out, shipped off to a concentration camp or shot. There was no help for this. . . . We had to write; it was our only defense against despair. Besides, the whole country was sown with the seeds of conspiracy and an "underground state" did exist in reality, so why shouldn't an underground literature exist as well. Except for two or three Nazi propaganda organs, no books or magazines were printed in the language of the defeated nation. Nonetheless, the cultural life of the country refused to be stifled. Underground publications were mimeographed on the run or illegally printed in a small format that was easy to circulate. Many underground lectures and authors' evenings were organized. There were even underground presentations of plays. All this raised the morale of the beaten but still fighting nation.[37]

Implicit in this paragraph are the tensions, drives, compulsions, and impulsions at work in a community. Any com-

36 Edward W. Barrett, *op. cit.*, p. 70.
37 Czeslaw Milosz, *op. cit.*, p. 88.

munication to such a group obviously requires sympathetic understanding, an understanding that may be hard for a "sending culture" that has never experienced life under such circumstances to achieve, but an understanding, nevertheless, that must be sought if there is to be true communication.

When we reflect on the civilization of Asia and on our own civilization, we are aware of great differences; the nature of these differences self-evidently makes the process of communication one that is fraught with perils. In executing this task, an essential premise is the establishment of communication so that both we who send and they who receive will understand each other.

Communication with the Home Audience

It is not enough, however, to be able to talk understandingly to another people. We must be able to talk to each other. This is a delicate point, and it is one that has aroused much discussion in the past. In World War II Congress was insistent that there should be no propagandizing of Americans. Obviously the line between explaining the reasons behind, and the need for, American foreign policy and propagandizing for acceptance of those reasons is a fine one. On the face of it, it does not seem possible, however, to conduct political communication without taking into account the climate at home. The responsible officials of a government activity in this field would be guilty of omissions if they concentrated on their target and neglected a situation in their own country that prevented the accomplishment of their mission. Since, in a free society, public opinion is an essential support for political communication, it is equally essential for that public opinion, itself, to understand the meaning of policy decisions and the reasons for them.

In a free society, the established organs for this are the press and radio broadcastings. There is no need for an official propaganda agency to present government views directly to its own people. It would not be tolerated. The point to make is that while such an agency must be able to communicate to a target beyond its own boundaries, at the same time it should not forget the need for communicating with its own people; it must do this through the independent organs set up for this purpose in its own society. There is here a valuable check on what is being communicated. If it squares with known facts and admitted national objectives, it will be given its hearing and, according to the authority with which it is proclaimed, may be expected to wield influence. If it is partisan, one-sided, palpably political, it will be legitimate material for criticism and opposition. But communication within its own culture must take place as clearly as it must be established with a foreign culture. The officials of a government agency of political communication have as much responsibility for keeping this factor in mind as they have for seeing to it that it is kept in mind in connection with the words their staff puts out for overseas consumption. This communication attempt does not have to be undertaken by political communication officials. In fact it is obviously better if it is done by the proper officials directing the various departments of government. But if it is not being done, it should be clear that the political communication officials are required to see that corrections are made in order that their mission overseas can gain maximum effect. This is one of the principal arguments supporting the position that the leading official in political communication should rank and sit with the senior officers of the government.

Sherwood has given a graphic picture of what can happen when a government's communication abroad is not in phase with the situation at home.

> The battle in Congress over extension of Selective Service was disagreeably synchronized with the Atlantic Conference. . . . The President's proclamation of a state of unlimited national emergency did not in any way bind Congress to concede that the national security was imperiled. . . . The emotional appeal of families who wanted their "boys" back was overwhelming. Even more important was the shocking morale of the men themselves: they didn't know why they were in the Army, they were muttering and shouting about promises made to them of only one year of this useless subjection to the brass hats; and, worst of all, great numbers of them were training with broomsticks for guns and with trucks masquerading as tanks, which made the whole process seem a ridiculous waste of time. The war seemed more remote than ever from American soil, with the Japanese apparently bogged down in China and Hitler getting himself more and more involved in Russia.[38]

It does not require much imagination to picture the plight of a government trying to talk to, reassure, and enlist the support of alien cultures when what it is seeking to communicate patently is not understood in its own society. It is a fair question: How can a government expect to put over its propaganda if, at the same time, it does not seem able to tell its own men what they are fighting for, and why? How, if it cannot communicate with its own people, whom it knows, with whose traditions, principles, and inherited wisdom it should be instinct—how can it hope to convince an alien society? The cultivation of the minds in the sending culture is essential to the desired action responses in the receiving culture.

Efficient Organization Required

A further premise on which political communication undertaken by a government squarely rests is that it must be

[38] Robert E. Sherwood, *Roosevelt and Hopkins: An Intimate History* (New York: Harper & Brothers, 1948), p. 366.

organized. Again, it is probable that no one will seriously challenge this premise. The fact remains, however, that it was not organized in World War I. It was not organized in World War II. It was not organized in the Korean war. It still is not organized, although much has been done, and we are in a better position than before the three previous wars.

Thomson's blistering words on political communication in World War II do not seem at all intemperate after studying the literature and history, and, furthermore, they reflect a general opinion:

> Wrongly conceived at the start, poorly led by the central political leaders, prevented from reconstitution on sound lines by Congressional action, torn by internal disagreement over fundamental purpose and methods, the performance of the information service during its first two years of operation is hardly characteristic of what could have been done under favorable circumstances.[39]

Political communication by the most powerful nation in the world, bearing the responsibility for leadership among the free countries, cannot be a minor operation. It is an example of all or nothing. The decision has already been taken and implemented to make it a major instrument of government. The logic of this decision, given the range and scope of the weapon, requires considerable attention to the manner of its organization. For one thing, it is too late to organize the military aspects of this activity once a hot war starts.[40]

39 Charles A. H. Thomson, *Overseas Information Service of the United States Government* (Washington, D. C.: The Brookings Institution, December, 1948), p. 365.

40 "Different ways of demoralizing or deceiving the enemy in wartime cannot be discovered without much travail when the propaganda machine is already functioning, when undeniable situations exist, and when military security and political policy circumscribe the frame of reference in which imagination is allowed to function." Leonard W. Doob, *Public Opinion and Propaganda*, p. 266.

But it is not only in terms of hot war that the argument for organization

On two grounds there is a strong case for organizing the function. In a hot war there is not the time. Prior to war, for it to exert its maximum effect, it must be integrated. Both these reasons support giving the problem immediate attention in terms of a continuing operation. As Barrett says, "If one assumes that the United States will face a difficult world contest for years, he must also assume that international persuasion will be a problem for a long time."[41] A problem of this magnitude obviously cannot be solved by piecemeal measures, patched together on a makeshift basis. In part, the evident reluctance to come to grips with this problem may have been occasioned by our dislike of attempting to manipulate people's thinking. If political communication is regarded as a dishonorable occupation, this could well stand as an excuse for pushing the problem aside except when it intruded itself unbearably. But for a long-term contest, a free society cannot afford to take this view. The activity is important and an accepted instrument of policy. The only alternative is to undertake it on honorable terms. In this context, it should be thoroughly organized and its terms of reference with relation to older agencies of government worked out and clarified. This cannot be done in time of emergency.

Pettee has pointed out:

> Collectively the nations associated with us in resistance to Communism, and in determination to survive in freedom, have still two-thirds of the manpower of the world.

is pressing. "For both the short term and the long term, then, first-rate strategy and mechanism for informing and persuading others become essential parts of government. These can be developed and continually improved at a cost that is minute beside other international expenditures. They can be fully effective, however, only if they operate as an integral part of a much larger concerted strategy." Edward W. Barrett, *op. cit.*, p. 226.

[41] Edward W. Barrett, *op. cit.*, p. 286.

> It is silly to suggest that we should turn our back on such human assets, aside from all questions of human values. But the methods by which to aid our allies, to strengthen them . . . to mobilize their will and spirit . . . is a problem we have only begun to master.[42]

A mobilization of the spirit is a task for political communication. It cannot be undertaken in the absence of proper organization.

Implicit in organization is knowing what we want to organize *for.* Political communication can be organized in a number of ways. The clue to which is the best way is given by deciding what ends are held in view, remembering always that guns can be silenced and planes grounded, but that words and ideas, once uttered, cannot be recalled.[43] As Norman Corwin said in one of his broadcasts: "Here's one last thought to fasten in your throat: who knows who may be listening, and where!"

All the evidence points to our having to undertake a program of political communication for a considerable period of years. We may not escape organizing it. The question is: How well are we going to do the job? The fact is: It is not yet done. And this in spite of the evidence of three wars.

42 George S. Pettee, The Walgreen Lectures, "The Soldiers and the Republic," Lecture V: "The Big Questions." University of Chicago, January, 1954. Unpublished manuscript.

43 ". . . in modern war propaganda is an offensive and defensive military arm which should stand on equality with that of the Service departments. . . . While tactical victories may be won by tricks and perversions of the truth, strategy demands a long-distance and coherent war policy, into the framework of which all propaganda should fall. . . . propaganda must be based on a long-term calculation of realities; . . . it must be consistent with geographic, economic and psychological facts; . . . whatever propaganda you make for the war will dog your steps in the peace and after it." Kingsley Martin, *Propaganda's Harvest* (London: Kegan Paul, 1941) (The Democratic Order, No. 13), p. 55.

We have seen the truth of this in Britain's propaganda to us in World War I under Northcliffe, and in the continuing effect of Japan's slogan, "Asia for the Asiatics," to give only two examples.

There is no need to look further than the daily press to count the examples that underscore our present lack of organization.[44]

The Significance of Professionalism

Unless political communication is organized, respected, and played by rules, it is useless to expect it to get results—at least the sort of results we want. It is self-evident that this means it must be a professional job. Furthermore, it is a job calling for experienced men, not inexperienced youths. This is not to discount the creative element. Nevertheless, the machine is important and it must be run by professionals. Schramm says "the world has progressed to a state in which self-preservation alone demands the most intense psywar pressure that a large body of trained professionals commanding immense resources can bring to bear."[45] It is Barrett's conviction that "There is no substitute for a strong mechanism staffed by trained specialists whose skills cannot be acquired overnight."[46]

Sherwood gives us a glimpse of the importance Roosevelt attached to professionalism. Political communication was not a hit-or-miss venture in his hands. "He studied every implication for its effect on various groups in the nation and on allies and enemies and neutrals."[47] Linebarger says propaganda "can become true psychological warfare, scientific in spirit and developed as a teachable skill, only by having its premises clearly stated, its mission defined, its instruments put in systematic readiness, and its operations

[44] See Ch. I, footnote 9.

[45] Wilbur Schramm, *et al.*, *The Nature of Psychological Warfare* (Baltimore: Operations Research Office, The Johns Hopkins University, 1953).

[46] Edward W. Barrett, *op. cit.*, pp. 299-300.

[47] Robert E. Sherwood, *op. cit.*, p. 217.

subject to at least partial check, only by the use of techniques borrowed from science."[48]

This requires professional competence. There is, at the moment, little training for political communication. Practitioners are pitched in to sink or swim. They learn by the process of immersion. There are good grounds for requiring practitioners in this field to have a solid background of professional experience. There are strong arguments for contending that training in objective reporting can bring greater understanding of both how and why it may be necessary to edit the news. Similarly, training in the responsibilities of professional broadcasting can bring awareness of the reasons why propaganda may require an altered emphasis. And, since in addition to the practitioner, social scientists are essential to competent prosecution of the activity, men in these branches of science must have professional training and be able to demonstrate their competence.

Tactical and Strategic Operations

The last premise apparent from the practice of political communication is that if the objective is short range in character, the nature of the activity will be tactical; if it is long range, it will be strategic. This must be so for a number of reasons, chief of which is that the mind, ordinarily, is slow to change. "Men's ideas, beliefs, and loyalties—their nonmaterial culture—are frequently slower to be changed than are their material tools."[49]

The short-range objective, therefore, is a sort of one-shot affair. The stimuli selected should be chosen in terms of producing an immediate result, and, obviously, the result

48 Paul M. A. Linebarger, "Warfare Psychologically Waged," in Daniel Lerner (ed.), *Propaganda in War and Crisis,* Ch. 13, p. 270.

49 Robert S. Lynd, *op. cit.,* p. 59.

must be something that may reasonably be expected to follow. An example of this, in combat, is the surrender of the enemy in circumstances when continued fighting is hopeless but may inflict casualties on one's own troops. Under given conditions it may be reasonable to expect the enemy to surrender without further fighting if the right appeal is made. This has, in fact, happened on a number of occasions. There is nothing long range about this.

The long-range objectives of political communication necessarily involve political themes; political themes are of the essence. They work slowly, in the main.[50] With this in mind, obviously no experienced practitioner will undertake to try to influence men's minds in a desired direction if the time allowance is inadequate for the particular end desired. Past history permits the opinion that it would not be difficult to document instances where strategic objectives were undertaken in tactical terms. In other words, political communication was instructed to propagandize for immediate action with appeals and arguments that could not possibly be expected to get results in the time available. The premise may be written like this: If the objective is a short-range one, it requires tactical treatment; if it is a long-range objective, it will involve at least elements of the strategic. It is important to keep this distinction in mind and to pin down clearly, in all planning for the activity, what the objectives are. There is no question but that our knowledge of the human mind, while most inadequate, is more than suf-

[50] "Propaganda does aim . . . to 'change attitudes'. . . . But this is only the means by which propaganda gets to its ultimate aim of influencing behavior. . . . It is important to recognize that the behavioral consequences of modified expectations need not appear immediately. Indeed some of the most important consequences of any attitude-alterations appear only in the longer-range future. Propaganda adjusts the timing of the behavioral consequences it seeks to the short-range or long-range aims of its policy goals." Daniel Lerner, "Effective Propaganda: Conditions and Evaluation," in Daniel Lerner (ed.), *Propaganda in War and Crisis*, Ch. 21, p. 348.

ficient to tell us definitively, in most instances, which objectives require long-range treatment and which short-range. Time, money, and effort are only wasted in trying to get immediate behavioral action on issues that, if they are to be affected at all, can only be affected slowly.

The premises that have been discussed so far in this chapter, and that, it is argued, logically govern the activities of a free society in the field to which we have put the name political communication, may now be summarized. Only in terms of political communication does it seem that we can cover an area of such manifest scope. As has been said earlier, psychological warfare is an inaccurate term for so wide a field. The concept of political communication both allows more latitude and places the components of the activity in truer perspective.

Basic Premises Governing Political Communication Conducted By A Free Society

1. Political communication should be based on facts. The accuracy of these facts should be capable of independent verification over time. If proved wrong, error should be acknowledged. In all probability this is the course that will produce confidence and conviction and preserve those values in a free society commonly accepted as "truths."
   (a) Public opinion is likely to support such a premise.
   (b) The technique of the feint in a clearly tactical situation is not precluded, but caution is required if credibility is to be sustained.
2. The ideas that sustain and advance the conception of a democratic free world possess force and should be the main theme and focus of strategic political communication.

(a) The presentation and interpretation of policies and actions must be clearly related to this main theme.
(b) It must be assumed that these policies and actions themselves are undertaken in terms of this premise.

3. Political communication must be congruent with national policy and objectives, and with the plans for their attainment.
   This is implied in the second premise, above.
4. Political communication will be ineffectual unless it rests on the considered basic opinions and attitudes of its home audience.

   (a) It is frequently true that leadership and public discussion are required to crystallize and clarify opinion and attitudes on relevant issues. But it cannot be assumed that the public can be tricked and manipulated into opinion alien to basic values.
   (b) Without the clarification of opinion thus provided, it is unlikely that adequate action responses will be forthcoming from the target.

5. Political communication, divorced from positive action, is futile.
6. Thus political communication is always in the role of an auxiliary to action. It is never in the primary role.
7. Effective political communication requires two kinds of knowledge:

   (a) Thorough knowledge of the cultural, political, and economic patterns of specific target audiences;
   (b) Knowledge of national policies and plans.

8. Political communication must be undertaken in terms of the action responses desired from its targets.
9. Political communication is effective to the extent to which the sender and the receiver can each assume the action perspective of the other. At the very least the

sender must be able to assume the role of the receiver.

10. Political communication as an instrument of state calls for efficient organization and central direction at the top executive level in government.
11. The operations of a planned political communication program are effective to the extent that the operators are explicitly trained in the required content and methods.
12. The strategic and tactical aspect of political communication must be clearly distinguished in policy, administration, and practice. Tactical principles and techniques must be adapted to the immediate situational requirements. They should not negate the basic themes of strategic communications except under extreme conditions.

It is not suggested that these premises are definitive; rather they should be regarded as preliminary formulations extracted from examination of our experience with political communication so far. Further systematic examination of experience, together with relevant research, ought to permit the refinement of these premises into accepted, definite principles governing the activity. In the meantime, however, on the basis of present knowledge, it is suggested that these formulations should be the foundation on which a planned operation in this field proceeds.

If the premises set forth in this chapter are accepted, there are certain implications and corollaries that are implicit in them and follow from them. In the first place it is possible to formulate a basic hypothesis: If we have an idea, and action that is congruent with that idea, then we may conduct effective political communication. It does not, of course, follow that we shall conduct effective political communication. There are a number of factors that may prevent it—for example, lack of organization, no support by public opinion, lack of knowledge. But even with these, lacking the juxtaposition of the idea and the right action, we should not expect effective political communication.

If this hypothesis holds up, it leads to an axiom: Political communication without action in support of it is ineffective. Words are not enough. Our output of words to a target, together with our actions in seeking to attain an objective, must both convey the same meaning. The same rationale in our minds must govern both and be expressed in both. If this is true, then, the greater and more positive the action, the greater and more positive can be the political communication. This, in turn, implies that negative action loses any initiative. Therefore take no action that is not positive. In this connection it may be pointed out that there are times when silence may be positive. Under certain circumstances the effect of not speaking may be greater than the effect to be gained from talking. There are times, of course, when positive action is not wise. It seems important, in such circumstances, to consider a neutral attitude rather than a negative one.

It seems to follow from this discussion that, in the society of a free power, apart from any considerations establishing the political authority as the controlling factor in its administrative processes, the logic requires political communication to be directly under the control of this political authority, which is, in turn, answerable to public opinion.

There is a distinction to be drawn between political communication carried on by great political leaders and the same activity conducted as a routine operation. The activity as it is practiced by great leaders essentially is an art, governed by the gift of genius. Obviously this puts this aspect of the instrument beyond the capabilities of any agency for political communication. No group of men, however dedicated, can be expected to write and deliver such words as Lincoln used in his Gettysburg Address, binding up wounds, gathering together the threads of a garment covering the soul of a nation torn by fratricidal passions. No propaganda agency, "in time of 'the breaking of nations'" can summon "blood, sweat, toil, and tears," and promise that this alchemy will yield a people's "finest hour." This power is the gift of the gods, and without it a nation has no life. A people may be conscious of their history, of the valor of their past, of rendezvous kept with Destiny, but the fusion of all the forces with which a great and powerful nation is instinct and the description of them in terms that fire the hearts of men are not, commonly, within the terms of reference of a propaganda agency. This capability is the lonely right of great leaders, but when it flares, as Pettee has said, it lights up "the wall of premises" behind the backs of the men and women engaging in the routine operation.

It is not sensible to expect such an operation to come forth with a Declaration of Independence, a Gettysburg Address, an Areopagitica. But it is sensible to seek for it inspired direction, knowledge of its capabilities and limitations, an understanding of where it is weak and where its strengths lie. Furthermore, political communication, if undertaken by a free power, must be an honorable, open activity with no stigma attached to it if able, competent, and professional men are to be attracted to it, and if that power is concerned with its standing in the eyes of other nations.

This means that it cannot resort to trickery, deception, and rumor-mongering. Its case, in the last analysis, rests on the truest moral and political principles for which it stands as a nation. It is somewhat surprising to note that trickery in our propaganda apparently is considered perfectly legitimate; perhaps, from the contexts in which it appears, even smart. For example, Anne O'Hare McCormick has reported that Radio Free Europe taunts the Communists "in solemn interviews with fictitious characters invented by Radio Free Europe."[51] In fact, however, political communication is an instrument of national policy, and it must be governed in its applications by much the same sort of thinking with which national policy is developed.

The writer believes the evidence supports the conclusion that we are still not clear in our minds what political communication is all about. If once we can reach agreement on the premises that should govern it in a free society, we shall be well on the way to establishing the rules that should govern its practice, and much of the present confusion should be cleared up. At the close of World War II, the final Report of Operations of the 12th Army Group contained this sentence: "Psychological Warfare has a destructive impact. Public Relations a constructive impact; . . ." It is not necessary to argue that a destructive practice is one that cannot win us friends if it is applied to our allies.

In a free society, operating on the principles embodied in our national documents, it seems evident that any use of political communication must be in terms that square with such principles. In the period ahead, a period that we may well assume will be long drawn out, political communication will be important. There is a definite need to consider deeply and logically the terms on which we, as a free people, may undertake it.

The logic of political administration in a free society requires that the ultimate direction of political communication must be in civilian hands. In its military context, precisely as military weapons and military commands are subordinate to political authority, psychological warfare must be answerable to political authority. The tactical application in combat is uniquely a military command responsibility. The strategic application involves longer-range

[51] Anne O'Hare McCormick, "A Focal Point in the Battle of the Air-Waves," *New York Times*, August 22, 1953, editorial page.

considerations that may well conflict with immediate military decisions.

In time of war, as a practical matter, trickery, deception, and rumor-mongering—all the attributes of black propaganda, in other words—have a role to play. But this role is subordinate to the main function of political communication and even in a hot war may not usurp the privileges and functions of the main national operation, which is required to be an open one.

III

The Nature of Political Communication: THE OPERATING PRINCIPLES

There should be no need to argue the proposition that if we undertake a national program of political communication the activity ought to be well organized, well understood, and generally supported. Furthermore, its principles of operation must meet with wide acceptance. In fact, so far, there is scant evidence that these operating principles have been given sustained consideration by government. If we desire to use this instrument effectively, these principles should be set forth and win acceptance, for only on the logic and force of the position they define may we hope to find them acceptable as touchstones by which we may successfully guide our operations.

The Requirement for Status

There can be little doubt that political communication has become one of the four main instruments for furthering national policy in a modern world. If it is to function efficiently, it requires equivalent status with the three older, established branches of government to which we have been long accustomed—the diplomatic, the military, and the economic.

No single action supports the contention that this instrument has, in fact, become of sufficient importance to rank with the other three. A number of developments taken together, however, combine to support the position. In the first place there is the Information Act of 1948 designed "To promote the better understanding of the United States among the peoples of the world and to strengthen cooperative international relations." Considering our ingrained dislike of attempts to influence the thinking of others, and the record of sustained opposition by the Congress to a propaganda agency, the final acceptance and support of the weapon tends in the direction of a supposition that only because no other instrument appeared to offer its capabilities did it win reluctant approval.

An additional factor is provided by the long struggle between the Department of State and the Department of Defense for the control of this instrument. This struggle is as yet unresolved. The general public is by no means aware of the nature of this conflict, nor the grounds on which it has been fought, nor the bureaucratic frustrations that repeatedly have hampered the development of an effective operation.[1] In the main the arguments have centered around the vexed questions of jurisdiction and procedure in time both of hot war and so-called peace. There has been little concern with an attempt to determine the premises and operating principles that, once defined and accepted, automatically act as governors and point to the correct solutions of problems involving control.

It is not logical to suppose that two such powerful departments as State and Defense, each responsible to so great an extent for a nation's policy, would have contended as

1 For an excellent vignette of the difficulties faced when logical administrative decisions conflict with political realities, see Arthur Krock, "Key Office for Nixon Deferred by President," *New York Times,* Feb. 10, 1957, p. E3.

bitterly as the record shows they have for control of a secondary instrument of no particular significance. And it will be remembered that both Presidents since World War II have been persuaded of the necessity for an adviser in this field, attached to the White House staff.

Partly our concern may be attributed to the evidence of the effectiveness of this weapon in totalitarian hands. Hitler and Goebbels accomplished major strokes of policy with it, albeit on corrupt premises. In Russian hands its capabilities have been given weightier demonstration. For five years the North Koreans were subjected to an intensive indoctrination by Soviet agents. No one who has any experience of the Korean war will defend the position that the North Koreans were merely deceived. They were firmly grounded in the tenets of communism and believed them and in this belief fought tenaciously and hard against their brethren in the South.

The impact of technological advance also must be taken into account. No longer may the instruments of national policy be exercised compartmentally. Their effects reach into all spheres of our lives, and decision-making reflects this. Thus we have the creation of the National Security Council, "an essential link because so many decisions that now had to be made were a composite of military and political questions."[2] Increasingly, the National Security Council has come to recognize, as well, that to this "composite" must be added the psychological. And the Senate committee that studied our overseas information programs drew careful attention to this psychological aspect in pointing out that the meanings and interpretations given by foreign observers to the actions of Congress, the executive, and the

[2] Walter Millis (ed.), *The Forrestal Diaries* (New York: The Viking Press, October, 1951). The words are Forrestal's, but the reasoning is attributed to General Norstad, p. 315.

judiciary, as well as to the actions of groups of citizens, "are now of deep international significance."[3]

It is this same committee that says that "the prestige of the [United States Information] agency and its head is the first and foremost criterion of evaluating the impact of the propaganda point of view."[4]

Both the logic of the instrument and the existing political realities are pushing firmly in the direction of establishing political communication as an instrument that should rank equally with our diplomatic, military, and economic activities as essential to the maintenance of the state and the attainment of our national objectives. In spite of this, it is a matter of record that there is still no unanimity of opinion regarding the over-all mission of the United States information agencies. Differing points of view have contributed to the uncertainty and confusion among information personnel. This is a serious handicap to the development of a coordinated and purposeful program.

Not until status is both yielded and accepted may we expect this instrument to develop its full potentialities. Kingsley Martin's comment on British experience in World War II is apposite to our own. Discussing the appointment of Brendon Bracken as head of the government information activities succeeding Duff Cooper, Reith, and Macmillan, he said, "the new Minister does at least begin with the knowledge that his job is impossible as long as the ministry is subordinate to the Service departments."[5]

Equality of status is required and its necessity demonstrated by the progression of historical incidents in the field.

[3] Report of the Committee on Foreign Relations, Overseas Information Programs of the United States (Hickenlooper Report), Senate Report No. 406, Eighty-third Congress, First Session (1953), p. 32.

[4] *Ibid.*, p. 148.

[5] Kingsley Martin, *Propaganda's Harvest* (London: Kegan Paul, 1941) (The Democratic Order, No. 13), p. 40.

What the instrument has accomplished in German and Russian hands on corrupt premises is proof of its capabilities. At least equal success may be expected in the hands of a free power as in the hands of dictated government, for political communication is inherently neither good nor evil. It is amoral. It is the premises and operating principles governing its employment that determine its direction.

The Scope of the Instrument

It is the nature of political communication itself that determines the second principle, namely that it may not function under the direction of any single department of government, but must serve the national interest. It must represent the sum of what all departments of government are attempting in the interests of the nation as a whole. Its activities cut across all the lines of national endeavor. It has no boundary except the limitations of the human mind. Thus it is difficult to assign responsibility for it to any one department of government. If it is assigned to Defense, whose primary concern is war, it is likely, as its history shows, to be restricted in its application. If it is assigned to the diplomatic field, its role in war may be neglected. It may not be subordinated to economic measures, for, self-evidently, the minds of men, which are what political communication seeks to influence, encompass a variety of interests that can be influenced by means other than economic. Its scope has been outlined by the Senate committee that investigated our information services. It is the role played by the Congress, by the executive, by the judiciary, and by the people and organizations made up of the people. No single department of government may pretend to administer an activity that depends for its success on the interacting forces of the entire nation. Yet all these forces play a part in political communication. It is their co-ordination that is essential to the mission of the instrument.

This activity cannot be carried on in a vacuum. The explanation and interpretation of our ideas must be meaningfully related to the actions of government. And unless these actions support the logic of the ideas we proclaim, political communication will carry little weight.

This is much more than a program of information. And it is much more than a matter of strategic or tactical psychological warfare during hostilities. It requires taking into account the role of the nation as that role is expressed by its political decisions, its military decisions, its domestic climate of opinion, its day-to-day activities, and its long-term desires and aspirations. The history of our propaganda attempts shows clearly that we have not acted on this concept, that, in fact, we are still not prepared to act on it.

National Policy Must Be Reflected at All Times

Political communication has to be considered as a continual process. It cannot be turned on and off like a faucet. It goes on all the time, either as a purposeless amalgam of a nation's activities from which observers draw their own conclusions, or else as a powerful adjunct to the actions and policy of government reinforcing and interpreting the main stream and development of that nation's life. As a purposive program it sees to it that observers are provided with the necessary information to draw correct conclusions. This is not solely a matter for political direction or military direction. At times the emphasis may be one or the other, but only insofar as the national policy at a given moment is predominantly political or military. Delbruck's concept of the inseparable nature of war and politics applies with even greater force to political communication. "His study of the military institutions of the past had shown him, in every age, the intimate relationship of war and politics, and had taught him that military and political strategy must go hand-in-hand."[6]

This concept has not been always apparent in the history. Wallace Carroll recalls his meeting with General Eisenhower before the TORCH[7] landings:

> "I don't know much about psychological warfare," he [Eisenhower] began, "but I want to give it every chance. . . . I want one man from OWI to work on the propaganda plans for TORCH. But I want to make it clear"—here he slapped his desk—"that he must become a member of my staff with all the privileges and duties of an officer on my staff." This was the Eisenhower formula which was to have far more important consequences than its author could have seen at the start. Every one who came into his theater, whether military or civilian, American or British, had to forsake old allegiances and submit to the authority of the theater commander.[8]

It is important to note that the nature of political communication refuses to permit this single allegiance. The allegiance political communication owes, and must owe, is to the over-all national policy, not merely its military aspect, not merely its political aspect. This attribute is important and constitutes a major problem. Obviously its solution must be in terms of national objectives, not in terms of political or military policies.

Authorities in this field of political communication have clearly recognized the complexities provided by this attribute of the instrument, and they have not hesitated to fight

[6] Gordon A. Craig, "Delbruck: The Military Historian," in Edward Meade Earle (ed.), *Makers of Modern Strategy* (Princeton, N. J.: Princeton University Press, 1952), Ch. 11, p. 261.

[7] TORCH was the code name for the North African landings in World War II.

[8] Wallace Carroll, *Persuade or Perish* (Boston: Houghton Mifflin Co., The Riverside Press, 1948), p. 12.

for the right to resist departmental restrictions when necessary in the national interest. Bruce Lockhart, Britain's Director General of Information in World War II, in discussing Robert Sherwood's status as head of the overseas branch of the Office of War Information, wrote: ". . . on matters of the highest policy and in moments of crisis, notably when the American chiefs of staff wished to take over the control of all propaganda, Bob was able to go direct to the President, and on those occasions his intervention was decisive."[9]

The nature of political communication will not, and cannot, allow any one department to "take over control of all propaganda." In part this flows from its quality of continuity.

> . . . psychological action, like intelligence, and springing from intelligence, far from being solely a weapon of warfare, is a natural means of implementing peace. It is the surest, in fact the only, means of orienting the peoples of conquered nations to the acceptance and understanding of the principles of human existence and government for which we fought.[10]

President Truman was referring to this quality when, at the conclusion of the United Nations Conference in San Francisco, he said: "It is easier to remove tyrants and destroy concentration camps than it is to kill the ideas which gave them birth and strength. Victory on the battlefield was essential, but it was not enough."[11]

Repeatedly those with experience in the field of propaganda have drawn attention to the need for continuity of operations. Propaganda in war may not ignore plans for peace after war. Equally, in peace, the fact that war may

9 Sir Robert H. Bruce Lockhart, *Comes the Reckoning* (London: Putnam and Co., 1949), p. 236.

10 Captain Ellis M. Zacharias, USN, *Secret Missions: The Story of an Intelligence Officer* (New York: G. P. Putnam's Sons, 1946), p. 390.

11 President Truman at the conclusion of the U. N. Conference in San Francisco, quoted in the *Department of State Bulletin,* July 1, 1945, p. 5.

break out conditions any realistic attempt to influence the minds of men. It is this quality of continuity that imposes the principle that political communication must first serve the national interest, only secondarily the departmental.

While the logic of political communication, the experience of practitioners, and the conclusions of students solidly support the contention that this instrument functions as a continuous process that must reflect, first, our national objectives, this concept is not yet generally accepted. An influential body of opinion in the Department of Defense holds that in a military theater the commander's decision in this field, as in those others directly relating to war, should be final. Equally, the United States Information Agency has not been concerned with the problems it might be expected to face if war broke out. It has been concentrating on its peacetime role. And its transformation from an information agency in peace to a psychological warfare agency, inevitable in the event of war, has occupied little of its time and thought. The facts do not permit this distinction to be drawn. In C. D. Jackson's words the activity "is not exclusively a military or political function, but straddles both spheres."[12] At the time Mr. Jackson left his post as Special Assistant to the President for Psychological Warfare, he was of the opinion that "polarization between military and political thinking on psywar has not yet jelled."[13]

This was undoubtedly true then and remains true today. But this failure to "jell" now is due more to recalcitrance than to ignorance. Enough dependable information, experience, and knowledge are available to permit what has been a fluid situation to "jell." Political communication is a curiously intricate and continuous process requiring the utmost in teamwork if it is to be effective. The only final authority

[12] C. D. Jackson in a talk with the writer recorded as a Memorandum for the Record. April, 1954.

[13] *Ibid.*

is public opinion. And because of this quality of continuity, political communication can be undertaken effectively only in terms of the entire range of national policy.

The Requirement for Congruity

Experienced authorities in this field are unanimous in concluding that there should be a close and frequent interchange between the makers of national policy and those who will be responsible for explaining and interpreting the significance of the policy—in other words, the operators or practitioners of political communication. This interchange has been lacking in the past, and it is still lacking today. Our Middle Eastern policy for example, announced after British and French intervention in the Suez dispute, was not developed in partnership with the United States Information Agency. Yet on the shoulders of the Information Agency fell the burden of interpreting it. Its senior officials, responsible for its direction and leadership, require this partnership if political communication is to be fully effective. It is manifest nonsense to argue, as frequently has been the case in history, that the leaders of a government propaganda agency may not be trusted with sensitive and delicate intelligence. If they do not possess this discretion, they should be replaced by men of caliber in their profession who do possess it. But to tie their hands in advance and at the same time criticize them for ineffectiveness is preposterous. So far, to a not inconsiderable extent, this is precisely what we have been doing.

Forrestal, who reflected thoughtfully on the nature of our government and was experienced in its machinery, observed that "There is need for much greater exchange of information and knowledge of what government policy is, so that all hands can move in conformity therewith."[14] A

14 Walter Millis (ed.), *op. cit.*, p. 87.

propaganda agency may not make policy, but, equally, unless those responsible for policy permit the potentialities of propaganda to be realized, the agency can do little to support the achievement of our national objectives. Too often policy-making and propaganda do not go hand in hand, and the result is loss of initiative and a defensive position when, with full partnership, a positive approach could be made, and the fruits of responsible leadership more clearly exemplified to those who watch and weigh the kind of leadership we are giving in international affairs.[15]

There is an absolute requirement for policy and propaganda to be intimately linked and, moreover, linked in terms that do not permit what is a short-term objective to deny or confuse or obstruct the long-range aims. When this is not the case, there is trouble, and there are plenty of examples in our recent history of the use of propaganda to support this—for example, the Berlin airlift. Here was a weapon of distinct psychological advantage as well as a weapon for the defeat of Russian obstructionism. But our propaganda machine, again, was not represented in the policy-making team, with the result that we were not able to announce what we could do with planes if necessary and then make the prediction good.

The impact of policy on propaganda has been well stated in the Hickenlooper Report. "The Voice of America . . . does not and should not make policy. If, as has often been true, policy decisions are uncertain, or laggard or confused, the corollary can only be a dilute, delayed, or bewildered Voice over the air waves."[16]

It is not always possible, for obvious reasons, to provide

[15] In this connection, see the Hickenlooper Report of the Senate committee on our overseas programs, previously cited, and the Summary of the Drogheda Report. See also the writings of Hans Speier on psychological warfare.

[16] Hickenlooper Report, p. 20.

propagandists with clean-cut policy. But when this is the case, it should be a deliberate action, not, as the record clearly shows has so often been the case, the result of carelessness. In the Korean war, for example, the policies laid down reflected the difficulties imposed on psychological warfare policy in the course of any war of limited and irregular character, in which the major powers were avoiding any overcommitment in their diplomacy. Nevertheless in the majority of instances there need be little difficulty in linking policy and propaganda. Both Germany and Russia —Russia especially—have been well aware of this linkage and have capitalized on it and exploited it. With all this evidence staring us in the face we have still not taken the necessary steps to see to it that our actions and our words complement and reinforce each other, and we have weakened our position in foreign eyes as a consequence. When, at the close of the Korean war, we refused to save the lives of our own men at the expense of the principle that prisoners could not be forced to return to dictated government against their will, we stood, undeniably, on a lofty moral rock. The action was impressive. It would have been more impressive still if, with modesty and firmness and intent, we had proclaimed the convictions on which this action was based. Instead, we left it to others to discuss in terms of their own predilections. We did not interpret and explain its significance. And it possessed a significance of which we could be justly and modestly proud.

National Policy the Controlling Factor

The linkage between policy and propaganda requires that our political communication be planned and co-ordinated, not in terms of specific diplomatic or military policies, but in terms of over-all national objectives. The indications seem beyond dispute. This has not been the case. There has not been effective central direction, there has not been planning and co-ordination. On rare occasions when these have been in evidence, there have been results.[17] In the main, however, the history of what we have called psychological warfare points to failure in co-ordination and planning, rather than success. In fact, we seem to have ignored the significance of co-ordination.[18]

Thomson voices the considered judgment of all thoughtful students of the subject when he says that "Any national information program should serve national interests and should be related to grand strategy. To give maximum service, the program should be kept constantly co-ordinated with all major elements of national action; political, military and economic, long-run and short-run."[19]

Those whose memory goes back to World War II will re-

17 ". . . the brilliant program by which the Navy, the State Department and the OWI collaborated to give the Japanese a definition of Unconditional Surrender which played an important part in hastening surrender." Charles A. H. Thomson, *Overseas Information Service of the United States Government* (Washington, D. C.: The Brookings Institution, December, 1948), p. 23.

18 "The failure of our leaders to sense the world propaganda implications of important actions has deprived us of the possible fruits of skillful exploitation of the work and statements of the Atomic Energy Commission, of such agencies as the National Security Resources Board, and other groups in a position to establish dramatically the United States position on topics of the keenest interest to foreign powers. Our distrust of government publicity has spilled over into these fields: the information policy specialist is confined by notions about his role to only the most discreet suggestions as to how such public statements and actions might be shaped to serve propaganda purposes." *Ibid.*, pp. 360-61.

19 *Ibid.*, p. 367. Note also Wilbur Schramm: "This matter of integrating psywar into the total international operations of a nation is the most important operations problem that psywar faces. It was not solved in World War II . . . and almost certainly has not been solved since. Yet until psywar is integrated into planning at the highest levels the full potentialities of the weapon will never be exploited." *The Nature of Psychological Warfare* (Baltimore: Operations Research Office, The Johns Hopkins University, 1953).

member the multiplicity of agencies concerned with propaganda; the Foreign Information Service, the Office of Facts and Figures, the Coordinator of Information for Inter-American Affairs, the Office of Strategic Services. In this instance planning was not only unco-ordinated but even proceeded independently of national policy decisions. In the Korean war it was the same. Timely common guidance to all agencies concerned with propaganda was not forthcoming. There was a lack of specific national propaganda directives. In part, as previously noted in connection with the Korean war, this resulted from the desire of the Allies to avoid overcommitment. But also, in part, it resulted from failure to implement the linkage between propaganda and national policy. In the case of the North Koreans precisely the reverse is to be observed. The enemy had planned and he was ready. With him, in Kingsley Martin's words, policy and propaganda were one, and they were co-ordinated.

Political Communication Must Be Represented in the Formulation of Policy

In order for the linkage between policy and propaganda to be effective, the propagandist must take part in the formulation of policy. Successful propaganda is based on the predication that the propagandist possesses full knowledge of what the national objectives are and of what the departments of government are doing to implement those objectives. The history of this subject shows that there has been continuous and strong resistance to this proposition. The literature shows just as clearly that on the part of both students and practitioners there is striking unanimity in concluding that political communication cannot be waged to its utmost effectiveness without such participation.[20] We

[20] Edward P. Lilly, Hans Speier, James P. Warburg, and Rear Admiral Ellis M. Zacharias, to name only a few, have all commented strongly on this point.

may recall the incident that opened this book—the dropping of the first atomic bomb—and pause to reflect on Thomson's words:

> The war role of the propagandist was humble, even though the volume of propaganda was stupendous. He was not called to policy councils to contribute to policy decisions from his special resources. He often was not told of important developments long enough in advance for him even to carry out the mechanical operations of propaganda of "news" efficiently and expeditiously. It was hardly a surprise to the propagandist that he should have had no opportunity whatsoever to help central leaders estimate the effects on world opinion of the atomic bomb. Even in such a revolutionary case, the leaders relied on their own rule-of-thumb estimates which, however reliable for judging United States opinion, were certainly not adequate to cover the complexities of world reaction. In consequence our propagandists faced the greatest strategic crisis of our history with no forewarning and no chance to plan what should be done to maximize favorable effects, and offset bad ones, nor was there any plan offered by the leaders.[21]

It was Lovett's opinion "that foreign policy, like a war plan, could not be stated in a single speech . . . but was a planned, continuing flow of actions and statements directed toward a central objective."[22] Unless the propagandist is himself a part of this "planned, continuing flow," he can scarcely, in the words of Bruce Lockhart, "be supple with an instinctive feeling" for the most effective approach to his target.

It is Speier's judgment that the directors of this activity "must know the foreign policy objectives of their country."[23]

21 Charles A. H. Thomson, *op. cit.*, pp. 364-65.

22 Walter Millis (ed.), *op. cit.*, p. 443.

23 Hans Speier, "Psychological Warfare Reconsidered," in Daniel Lerner (ed.), *Propaganda in War and Crisis* (New York: George W. Stewart, Inc., 1951), Ch. 27, p. 479.

These objectives do not become known from hand-outs of official statements already in final form. The "continuing flow of actions" that culminates in policy may only really be known if, in the process of its formulation, there is participation. It is this participation that brings the suppleness of instinctive feeling. An activity that is neither art nor science, but compounded of elements of both, calls for this suppleness, and masters of it have possessed this suppleness; witness, for example, Franklin D. Roosevelt and Sir Winston Churchill.[24]

Any careful student of the history and literature of this subject, and of the personal experiences of responsible officials who have been engaged in this activity, will not fail to note the repeated insistence with which it is put forward that effective propaganda requires participation by the propagandist in the formulation of policy. The requirement is one that emerges with clarity and unanimity.

Creative Leadership Essential

Political communication requires much more than merely the ordering and administration of machinery. There is a

24 Lockhart holds the opinion that "political warfare is and always must be the handmaid of official policy. It is the duty and function of the propagandist to use his ingenuity, skill and special knowledge for the propaganda exploitation of official policy." Sir Robert H. Bruce Lockhart, K.C.M.G., "Political Warfare," *Journal of the Royal United Service Institution* (London, Whitehall), May, 1950, p. 201.

Note also the conclusion of the Drogheda Committee: "The first requirement is that Her Majesty's Government's policy should be clearly stated by Ministers at the time and in the form required to impress public opinion overseas. This means that the points of view of the Heads of Information Departments in London should be taken into account at the time when policy is being formulated and ensure that the requirements of overseas propaganda will not be overlooked. We are by no means satisfied that this always happens at present." Summary of the Report of the Independent Committee of Enquiry into the Overseas Information Services, April, 1954, Miscellaneous No. 12 (1954), Her Majesty's Stationery Office, London, p. 20.

requirement for leadership and, moreover, creative leadership. In Sherwood's words, concerning the situation early in World War II, "Spiritual preparedness for coping with powerful evil was required before it began to occur to people that some tanks and bombers and light aircraft carriers might also be helpful."[25]

This "spiritual preparedness" is the essence of propaganda conducted by a free power. The presence of this quality is not, primarily, a military or political responsibility. Essentially it is an expression of the quality that makes up the heart of a nation. It is intangible. When it is present, it is discerned by friends, and its existence is known to enemies. It is an ingredient in the moral stature of a nation. It may be either good or evil, depending on the definitions selected for those words. There is no question, for example, but that Japan was spiritually prepared for war. In our eyes it was a daemonic preparation. Just as surely, given our own ethical concepts, the quality of this spiritual preparedness will be evident if it has been nurtured. It is an important element of propaganda. It calls for creative leadership if it is to be effectively displayed. Sir Winston Churchill possessed the gift of being able to summon it for his people. Woodrow Wilson summoned it for the peoples of the world. When it is in evidence, it is a powerful force, as any history of the human spirit shows.

> It is a mistaken view of history to think that the responsibility of world leadership is something that has been thrust on our unwilling brow. . . . But if we are with due humility to live up to this responsibility, our propaganda cannot be merely an instrument of defense or offense. It must be transmuted by creative leadership into a vehicle of statesmanship. If this can be done, propaganda may be returned to its original meaning—the propagation of faith, and propaganda will then be seen not in a narrow

[25] Robert E. Sherwood, *Roosevelt and Hopkins: An Intimate History* (New York: Harper & Brothers, 1948), p. xiii.

> chauvinistic sense but as a means of continuing progress toward our ideal of the free dignity of man. In this sense, American government propaganda in peace or in war must be guided by people who are clear about the essential meaning of American life.[26]

If we are to propagandize not for an ideology—the ideology of the American way of life—but, instead, for the truth of our principles and convictions as a nation, leadership of this kind is indispensable to our propaganda. If, on the other hand, we regard the present situation merely as a conflict of one ideology—our own—in conflict with another powerful ideology—communism, we may expect to wind up inevitably in conflict. We are fighting an ideology, unquestionably. We can fight it with a better ideology. We can fight it best, as a free people, by demonstrating the truth of our principles. In no other way can we expect to overcome it so decisively that we win to our own side those peoples still wavering in their allegiances or convince those presently committed to the ideas we believe false. To accomplish this requires high qualities of creative leadership, the sort of leadership that Lincoln gave in saving the Union and thereby establishing before the world the intangible quality bound up in the name *United States*.[27]

The task of political communication is not merely to present an argument with wit and skill. It is to give intellectual and spiritual significance to our concern with the material

26 Ralph Block, "Propaganda and the Free Society," *Public Opinion Quarterly*, Winter, 1948-49.

27 A paper prepared under the auspices of Chatham House, London, puts it this way: "Defence in peacetime . . . involves the daily life of peoples, the news and ideas that reach them, the things they can buy and the taxes they pay; there is no major aspect of domestic policy that does not affect a nation's capacity to survive in the political and ideological struggle now being waged . . . high morale and an intelligent belief in national policy as a way of defending certain values and traditions are no less important than military and economic strength." *Defence in the Cold War* (London: Royal Institute of International Affairs, 1950), p. 30.

requirements necessary for a people's well-being.[28] The weapon for this is propaganda, propaganda in the original meaning of the word. Effective propaganda must be backed by political, military, and moral power. The use of these powers to support, in Malik's words, "the positive values of the western tradition" is the prerogative of the statesman. The convincing of others of the validity of this tradition is the prerogative of the propagandist, and this calls for much more than the setting up of machinery to spew out words. It calls for creative thought and creative leadership.

Specialized Machinery Required

Machinery is necessary, nevertheless, since the sort of leadership that is essential cannot be exercised without organization. For some ten years now we have been attempting to create such an organization. The fact that we have not yet succeeded indicates that it may be instructive to examine the requirements that demand peculiar means for the achievement of ends.

In the first place, in a free society that is also bureaucratic, any new organ of the state probably faces opposition. Established agencies and departments appear to fear such an organ as possibly encroaching on what they have come to consider as their preserves. Almost instinctively, therefore, as does the hedgehog when he is poked, they assume a defensive rather than a co-operative posture. In this atmosphere a new agency, fighting for its existence, appears to become jealous of its insecure position and tends to approach its problems not from the point of view of the logic of its mission, but from the viewpoint of sheer existence.

[28] As Dr. Charles Malik, of Lebanon, has expressed it: "In the end everything is a function of the mind and the spirit. . . . But the greatest need is to develop a grounded faith in the positive values of the western tradition. . . . Now when the West concretely and responsibly realizes what it is heir to it will have an infinite faith in itself." Address to government officials, Washington, D. C., December, 1952.

A propaganda operation conducted by a powerful state in the twentieth century, with all the technological tools it has at its disposal for getting its messages to big audiences, possesses powerful potentialities. It is natural for this power to be both feared and courted by the elite of such a state. If it is not under at least the partial control of departments they administer, they are fearful. If there is a possibility of acquiring this power, they court it.

Because of the nature of political communication, it cannot, logically, fall into any single department of government. To create a separate agency and give it the machinery for fully effective operation will tend to concentrate power, and a free society, out of experience, resists.[29] In other words, a single propaganda agency with authority to co-ordinate government propaganda is too powerful. If there is not this co-ordination, it is ineffective.

29 The struggle for power which is continuously going on in the capitals of nations, plus the nature of political relationships, makes very difficult the logical and orderly distribution of authority. Sherwood, who saw Washington at first hand, in discussing Hopkins' relinquishing the responsibilities for Lend-Lease to Stettinius, remarks: "Here he was calmly relinquishing an imposing title and a job of vast importance—and this in the city of Washington where men in top positions not only fought and bled to hold on to their own jobs but sat up nights figuring ways to take powers and functions away from others of equal eminence." Robert E. Sherwood, *op. cit.*, p. 376.

The Economist, discussing the Report of the Drogheda Committee, describes a situation uncomfortably similar to our own. "It is not a pretty picture that the Drogheda Report sketches, with its references to an interdepartmental committee fighting to share out a dwindling income, to the lack of any central plan or principles for information policy, to the division of responsibility among ministers, to the contest between jobs that are expected of the information services and the tools with which they have to work. . . . It looks, indeed, as if the direction of British overseas information services lacks not only coherence and central guidance but also the imagination to see what opportunities the future will bring. . . . The Government will be well advised to see that information work is kept henceforth under continuous scrutiny by critical and independent people, who ask awkward questions and lift policy-making above rivalries between departments." "Cinderella in Downing Street," *The Economist,* London, May 8, 1954.

No other department of government faces quite this problem, simply because no other department of government is quite so interdependent as propaganda is required to be. This is a direct result of its comprehensive nature. One of the difficulties, therefore, is to provide for co-ordination while, at the same time, limiting power. This is a peculiar requirement that other older and more established agencies of government have not had to face in the same degree.

Because it is a continuous process, propaganda cannot be transferred from department to department depending on circumstances of war or peace. Propaganda is concerned with overt action that takes place as a result of changed attitudes. The changing of attitudes is almost always a long, slow process. The work that may have been begun when a nation was conducting its business by diplomacy will almost certainly need to be continued when it has had to resort to war. If propaganda is to be successful, it will be necessary to continue the pressure on the target, to increase it, multiply it, but certainly not turn the task over to new operators unfamiliar with what is past. If this is correct, it means that the military and diplomatic arts must marry as they have never married in the past.

It is also apparently necessary for a propaganda organization to be in close touch with all who serve it and with its tools. It cannot operate by remote control.[30] It has already

[30] Lockhart recalls General Eisenhower's placing the "Anglo-American [psychological warfare] organizations under one head, the arrangement being that the military political warfare branch should conduct all operations in the field and accept political guidance from the civil agencies in London. Once again geographical separation might have created serious difficulties, but General McClure—the American officer in charge of the military political warfare branch, was not only fair-minded but essentially tactful. Largely for this reason a complicated machine worked reasonably well." Sir Robert H. Bruce Lockhart, K.C.M.G., "Political Warfare," *Journal of the Royal United Service Institution* (London, Whitehall), May, 1950, p. 197.

Note also Lockhart's opinion that "control of broadcasting cannot be

been pointed out that it cannot serve solely military policy at one time and solely political policy at another. The history of this activity shows its lack of congruence with national objectives when this is not taken into account.[31] What is required is an organization that will permit propaganda to serve several masters at the same time, something we have always been led to believe is impossible.

In the past, and in all likelihood as it will be in the future unless adequate preparation is made, the severest strains have been imposed when the direction of psychological warfare in a military theater of operations has not been congruent with the central direction of the activity in Washington. According to a tradition which is at least three centuries old, the influence of the military upon politics is a very dangerous thing in a constitutional democracy. If theater control of propaganda warfare is not precisely politics, it is, nevertheless, treading closely on its heels. "Duty to the State alone avoids false implications, and carries the appropriate implication of obedience to proper civil authority as the responsible expression of the authority of the State."[32]

Duty to the state is clearly the commanding tenet for

exercised efficiently unless the controlling body itself is close to the microphones." *Ibid.*, p. 194.

Our own Foreign Information Service in World War II had an operating office in New York supposedly under the direction of Washington headquarters. The control was imperfect. More recently the Voice of America moved its studios from New York to Washington, where it is directly under the thumb of the central office.

31 "In general, it may be said that the soldiers in the theatre of war regarded propaganda solely as an adjunct to military operations. . . . Inevitably the propagandists under General Eisenhower used all their ingenuity to assist the military. In London, OWI and PWE were the servants of their governments." Sir Robert H. Bruce Lockhart, *Comes the Reckoning* (London: Putnam and Co., 1949), p. 251.

32 George S. Pettee, The Walgreen Lectures, "The Soldiers and the Republic," Lecture II: "The Sense of Duty." University of Chicago, January, 1954. Unpublished manuscript.

propaganda, but the organization that will permit this and at the same time permit duty to be observed when a soldier is in command or when an ambassador is in charge is what so far we have not been able to accomplish. In default we have had situations such as the one that called forth a telegram from the Joint Chiefs of Staff to London "proposing that in view of the obvious confusion between London, Washington and General Eisenhower's headquarters, the combined chiefs of staff should take over the coordination of all propaganda policy. If accepted, the proposal meant that the virtual control of all propaganda to Europe would be exercised from Washington."[33]

In the absence of the development of any satisfactory organization, it must be obvious that whatever department of government has primary responsibility at a given time will attempt to secure control, for the sake of order. We face a requirement for developing distinctive machinery if we wish a really effective organization for political communication. That this will be difficult goes without saying. That it is impossible surely cannot be admitted. The problem is complicated by the present concern with the question of how national policy may be made and implemented most effectively in a period that promises to be one of continuing criticality. It is disturbing to the informed student to see that political communication is still being disregarded as possible revisions of machinery are debated. In spite of the history and the experience and the authoritative judgment of men familiar with the instrument, the tendency still appears to be in the direction of subordinating it to the more established tools with which a government seeks to gain its ends. And, if the lessons already in the record and repeatedly sustained mean anything, the only predictable result must be, as in the past, yet once more a confused, only par-

[33] Sir Robert H. Bruce Lockhart, *Comes the Reckoning,* pp. 253-54.

tially effective program when, with a little imagination, plus some executive decisions, a really significant instrument could be forged.

Lockhart, out of his experience as Director-General of Britain's wartime information activities, apparently favors in effect a czar for political communication.[34] It is doubtful if such a figure would be appropriate to the American landscape. As Forrestal says, "It is a pretty good thing for anyone charged with the responsibility of great decisions to have to maintain his point of view before an earnest and intelligent opposition."[35] Also, considering the scope of political communication, it is doubtful if any one man is wise enough to be charged with such a responsibility. The demands of modern society have so developed that it must appear beyond the capability of any one individual to attend to them. It is not possible any longer in war even to have a major decision made by one man.[36] The nature of political communication compounds this situation.

On the other hand there is unquestioned need for machinery that will co-ordinate and accomplish the task quickly and that, moreover, possesses the capability of resolving difficulties. At present no such machinery exists. The

34 "I recommended a separate department under the Foreign Office with close relations with the Chiefs of Staff organization and with a special junior minister in charge. . . . The Cabinet decided to give control to a committee of three Ministers. . . . A Committee of three officials with myself as chairman was to run the new Political Warfare Executive and to report weekly to the three Ministers. . . . A committee of three is two too many; two committees of three are an absurdity." Sir Robert H. Bruce Lockhart, K.C.M.G., "Political Warfare," *Journal of the Royal United Service Institution* (London, Whitehall), May, 1950, p. 194.

35 Walter Millis (ed.), *op. cit.*, p. 92.

36 "The General Staff, as now expanded in sheer mental capability for examining data and solving problems, has outgrown the capacity of the single mind of any commander to receive and judge the advice that can be given." George S. Pettee, The Walgreen Lectures, "The Soldiers and the Republic," Lecture VI: "Responsibility." University of Chicago, January, 1954. Unpublished manuscript.

Psychological Strategy Board, which was designed to correct this situation, found itself incapable of doing it.[37] The Operations Coordinating Board, which replaced the Psychological Strategy Board, has not been able to do it. At present, if operating agencies of government in this field persist in intransigent stands on questions of jurisdiction and control in this field, there is no machinery that can resolve their differences short of a direct appeal to the President for personal action and a command decision.

There is one further aspect of the problem of an adequate organization for the conduct of political communication. The logic of an effective operation in this field calls for facing the necessity to communicate with our own people. There has been studied and relentless opposition to the concept of the government's "propagandizing" its own citizens. There can be no quarrel with this position. But it is not a question of propaganda. It is a question of the information that public opinion requires if it is to act in accord with existing political realities. In Irion's words: "Failure to differentiate between an appeal on one side and propaganda and indoctrination on the other has caused considerable difficulty in determining policy on how information should be conveyed to the people."[38]

Tactical Psychological Warfare

Because of past experience, it needs to be clearly stated that since all phases of political communication are linked and interdependent, this includes even tactical psychological warfare. Strategic psychological warfare, of course, is

[37] "Observations . . . suggest that the component agencies of the [Psychological Strategy] Board . . . are continuing to go their separate ways in matters of psychological policy . . . this country cannot afford to have conflicting policies." Hickenlooper Report, p. 6.

[38] Frederick C. Irion, *Public Opinion and Propaganda* (New York: Thomas Y. Crowell Co., 1950), p. 9.

the proper name for that phase of political communication directed against an enemy after the outbreak of hostilities. As such, it is self-evidently directly related to, and concerned with, all activities in this field directed as well to allies and neutrals.

The relationship of tactical psychological warfare, however, is not always so apparent. For one thing, generally speaking, this phase is concerned with the enemy at the front line. The aim, for example, may well be the surrender of a specific enemy unit. In such a circumstance the operation may be carried out by soldiers with loudspeakers, and, seemingly, it may have little connection with the longer-range strategic aspects. The job is to get the enemy to default without further fighting, and once this is done the operation is finished. It looks very much as if it were, or could be, a tidy and separate little bit of work that could be carried out on independent terms without any reference to the daily activities of the strategic operation.

In fact, it is separate. But it is not independent. In the first place it may not negate or disregard the longer-range plans. It must be consonant with such plans, and, except in specialized circumstances, must conform to the letter and the spirit of the basic premises governing political communication as a whole. For instance, if the strategic mission prohibits concessions to an enemy target as a group, isolated units in that group may not be tricked into acceptance of a propaganda appeal by offering such a concession as a special case.

This is only another way of saying that, when it comes to fighting, the armed forces of a nation have to be given a relatively free hand. National ethics and national policy, however, still take precedence. Tactical psychological warfare is a problem of military concern and must be conducted under military direction. It is still related to the long-range aims and objectives of political communication.

Intelligence

The problem of intelligence for political communication will be discussed in some detail in a later chapter. It is necessary to draw attention to it here because the necessity for intelligence is a cardinal operating principle. Political communication makes special and peculiar demands on intelligence, and the problem is closely linked to that of organization. Lerner has succinctly shown the nature of the relationship:

> One requirement of sound policy is the *clarification* of goals; a second is their *instrumentation.* The two interact incessantly in the course of political life. . . . The step from clarification of goals to their instrumentation requires, in the first instance, adequate intelligence of the conditions which are "given." This crucial service rendered to policy-making by the intelligence function is perhaps the most serious area of ignorance in our present understanding of the political process.[39]

The requirement of intelligence for political communication puts it in a special class, requires special machinery, and, at the same time, draws on all the conventional intelligence methods available.

> Policy defines the limits within which propaganda must work by specifying the goals to be achieved. Intelligence defines the limits within which propaganda must work by specifying the limits of what the audience will believe and what they will do. Propaganda strategy consists of making the most of the terrain between the boundaries defined by policy and intelligence.[40]

39 Daniel Lerner, "Introduction," in Daniel Lerner (ed.), *Propaganda in War and Crisis,* p. xiv.

40 Daniel Lerner, "Part II: Policy, Intelligence, and Propaganda Strategy," in Daniel Lerner (ed.), *Propaganda in War and Crisis,* p. 53.

Full Administration Support a Prerequisite

Finally, contemplation of our past history and experience with political communication seems to show that this instrument requires firm support at the highest political level of government. Without this support, our experience so far shows, its problems are buffeted among departments and for want of resolution lie stalemated until a situation arises when action is inescapable. Under such circumstances, on a number of occasions in the past, there has not been time to give careful thought and study to a solution. Action has been taken, but the roots of the problem have not been dealt with.

There are at present serious differences of opinion among government departments on the question of delineation of responsibilities and on jurisdiction in military theaters. The Operations Coordinating Board has not been able to resolve these differences, nor have even higher levels. The plain fact is that if hostilities broke out, and a jurisdictional dispute arose, it would require settlement on an *ad hoc* basis.

The collected experience of two world wars, the Korean conflict, our experience with the cold war, and the body of literature that has grown up around the subject of political communication permit a decision to be taken on the basis of cool, dispassionate examination of pertinent factors. An established instrument of national policy, which is what propaganda has become, requires such a studied decision before and not during an emergency.

Dr. Leonard S. Cottrell, Jr., who has had considerable experience with the problems of political communication, considers that "Psywar, in its broad conceptions, should be represented at the policy level. The man so chosen should not be one who works with it as an advertising gadget, but one who can contribute broad thinking. Representation should be at the level next to the President, for the way we define our case to the world is of sufficient importance to warrant this."[41]

The very nature of political communication, if it is to be attempted with maximum effect, calls for support at the highest level. If we are to try to change the minds of men and win them to support the principles for which as a nation we stand, it is self-evident that we must, among other things, make clear to them the reasons for the actions we take—in other words, develop and explain our policy. Only if this is supported at the top, can it be expected to affect attitudes. The critical decisions of the modern world can be made only at the highest levels. The spokesmen for those decisions must be at the same levels. Unless such matters are "dealt with on the highest level and unless some clear-cut policy is laid down, what political warfare did in the War and what it can do now is, in fact, very limited."[42]

If there is agreement with the conclusions that have been drawn from a study of past experience and history, then the preceding discussion may be summarized in the following terms, and these terms may be used as governors of a national propaganda machine against which projected policies, plans, and operations may be checked in an attempt to ensure the maximum effectiveness of an organization.

The Operating Principles

1. Political communication is one of the four main instruments for attaining national objectives in a modern

[41] Dr. Cottrell in a talk with a staff member of the Operations Research Office recorded as a Memorandum for the Record, June, 1954.

[42] Sir Robert H. Bruce Lockhart, K.C.M.G., "Political Warfare," *Journal of the Royal United Service Institution* (London, Whitehall), May, 1950, quote by Lockhart in report of Discussion following lecture, p. 205.

world. It requires coequal status with the other three: diplomatic, military, economic.

2. The mission of political communication is all-embracing. Its activities cut across all lines of human endeavor. Its only boundary is whatever limitations the human mind may possess.
3. Political communication is a continuous process in the modern world. It is not a military function at one time and a Department of State function at another. It is both, plus the over-all national policy, all the time.
4. Political communication must be conducted in terms of long-term as well as short-term planning. The entire range of national policy and national objectives must be taken into account.
5. Political communication must be planned and co-ordinated in terms of the national objectives; not in terms of the policies of a single service or department.
6. The propagandist must be a part of the policy formulation process. He must have full knowledge of what the national objectives are and of what the departments are doing to implement those objectives.
7. Political communication requires creative leadership.
8. Organized political communication requires peculiar machinery, distinct to itself, if it is to function as a national instrument.
9. Tactical psychological warfare is separate from strategic political communication, but the two support each other and must be linked.
10. Intelligence for political communication, like its organization, is a peculiar and specialized function.
11. Political communication must be supported at the highest level in the government structure of the nation.

IV Wartime Difficulties and Problems

There are three significant phases in the recent development of political communication as an instrument of state. The first is our experience in World War II. The second is our experience in the Korean war. The third is the development of the instrument as a weapon of the cold war.

World War II

We were starting from scratch in World War II when we attempted to organize a psychological warfare operation. It is true there had been some experience with the weapon in World War I under the Creel Committee, but little of this experience was appraised and analyzed at the end of that war. And when we went into World War II, we were, in reality, faced with learning all over again.

Not that there would have been too much to learn. Modern methods of communication were barely coming into their own when the first armistice was signed. In fact, radio broadcasting did not begin for another seven years. Nevertheless there were certain experiences and developments that, if we had pondered them, might have helped us to avoid some of the mistakes we made.

However, the basic fact that seems evident, on looking back, is that we did no homework in the twenty odd years between the wars. Then, when World War II came, we were too busy in all directions to be able to sit down and plan and organize intelligently. The decisions were *ad hoc* decisions. And the men who made them were, in a number of instances, unfamiliar with the instrument they had to use and the nature of its relationship to the already established instruments of state—military force, economic pressure, and diplomacy.

Military Reluctance To Accept the Weapon

One of the first things that strikes us is the reluctance of the military to accept psychological warfare. This was perfectly natural and even predictable on the basis of the ancient and time-tested truism that new weapons are always suspect. Psychological warfare was no exception. Those who took part in its operations in World War II had many different experiences, but there was one problem they all shared in common: a resistance to psychological warfare by military personnel not directly associated with it.

The men who were engaged in this activity were rarely if ever part of the team, and the documented record is very clear in showing that they had difficulty in getting the equipment they needed to work with, in obtaining anything like adequate intelligence, in getting their messages distributed, and in being able to operate in forward areas.

All of this could easily have been changed if the command structure had wished to change it. The fact is the command structure was not, itself, persuaded of the value of the instrument. And, at the same time, it had a major war on its hands, and it was busy enough running that war without taking time out to consider the perplexities a new kind of operation was constantly raising.

The main element of this weapon was news. And it is not difficult to understand military objections to a weapon that had, as its main component, the spreading of news, or comment on the news, by air, by leaflet, or by newspaper. Our allies, the British, faced the same difficulty, and Lockhart records that Air Marshal Sir Arthur Harris "objected strongly to exposing his bomber crews to danger in order to drop what he called 'pieces of bumph.' "[1]

It is easy to see now how much more there is to this weapon than merely the furnishing of news, but two decades ago we hardly knew what we were getting into when we undertook psychological warfare, and news looked then as though it was the main prop to lean on. It is still, of course, one of the main props, but we have learned that there is much more to the instrument than this. When we began, however, it was not only news we wanted to give out. It was news as the American press understands the term. That is, the importance of selecting, explaining, and interpreting that news was not always grasped, even by those engaged in putting it out.

It is not at all difficult, with hindsight, to understand why the military looked askance at what, on first sight, appeared to be a news operation to which they were asked to contribute equipment, intelligence, and logistical support.

The Struggle for Control

The building of mutual confidence and the creation of a reputation for reliability and discretion were made more

[1] Sir Robert H. Bruce Lockhart, *Comes the Reckoning* (London: Putnam and Co., 1949), p. 171.

difficult under these circumstances by the sight of the many different agencies in Washington all concerned with varying aspects of psychological warfare and, for the most part, all without any clear understanding of the necessity for order or any real appreciation of the tasks and the nature of the relationship that ought to exist between themselves and the established organs of government.

At one time the Army had a psychological warfare branch. The Office of the Coordinator of Information (COI) was operating under General Donovan. Robert Sherwood was running the Foreign Information Service (FIS). The Office of the Coordinator of Inter-American Affairs (CIAA) was functioning under Nelson Rockefeller. Archibald MacLeish was the director of the Office of Facts and Figures (OFF). Eventually the Office of War Information (OWI) marched onto the field and took over the business of domestic information and open psychological warfare while the Office of Strategic Services (OSS) under Donovan concentrated on the "black" operations of psychological warfare.

With such a hodgepodge, jurisdictional disputes, tests of power, and, on the part of the military, unwillingness to cooperate fully were inevitable. CIAA under Nelson Rockefeller, for example, considered itself responsible solely to the President. When the Department of State took the position that CIAA must not only inform it of what it proposed to do in South America, but also wait for State Department approval, CIAA demurred. The issue was carried to the President, who supported the Department of State.

General Donovan, according to all indications, wanted to take over all our psychological activities and proposed an office for the job. This was fought by OWI and again, on appeal to the President, the direction and control of propaganda and psychological warfare were retained in civilian hands.[2]

It is clearly implied that General Donovan wanted to run all government information activities. Sherwood wanted to move his sphere out from under Donovan. Rockefeller wanted to keep information to the Americas separate from our other propaganda activities. The departments of State, War, and Navy, watching this struggle, were anxious to get settled who was in fact to run overseas propaganda. And in addition the military departments were concerned with safeguarding military information. They were not impressed with the activity as one that could make their task easier. Furthermore, FIS, under Sherwood, did not want the Department of State to have any hand in operations.[3]

The details and nature of this struggle for control have been set forth by such writers as Warburg and Thomson. It is much to be hoped that before long students will be welcoming the proof of Dr. Edward P. Lilly's scholarship and erudition in this field.

It is important to bear in mind that the beginnings of political communication have been marked by this struggle and its accompanying tests of power, for the positions taken later by various agencies, and difficulties still facing us today, have their roots in this troubled period at the beginning of World War II.

[2] Warburg records: "By May of 1942, it became obvious that neither the Co-ordinator of Information, nor the Co-ordinator of Inter-American Affairs, nor the Office of Facts and Figures were working satisfactorily—largely because of the confusion and overlap in their mandates and because the work of the three agencies was uncoordinated by any common policy-making body." James P. Warburg, *Unwritten Treaty* (New York: Harcourt, Brace & Co., 1946), p. 82.

[3] See Charles A. H. Thomson, *Overseas Information Service of the United States Government* (Washington, D. C.: The Brookings Institution, December, 1948), pp. 17-18.

The Importance of Planning, Organization, and Direction

It is obvious that a planned propaganda program must first of all determine what it desires to accomplish; it must have an objective. Secondly, it must either own or control the tools with which it will transmit its messages. In both World War I and World War II these axioms had to be learned and relearned. In World War I, for example, the Creel Committee was unable to produce military leaflets. Therefore the military produced them although both Creel himself and President Wilson had misgivings. The decision taken then to let the military produce propaganda leaflets provided a precedent that was referred to when the issue arose again in World War II. This issue revolved around the question of policy. Who was to determine what was to be said in such leaflets and control them? The issue still has not been settled. And there are still conflicting views in the government on the question of how far civilian control of policy for psychological warfare should go.

One factor may well have been the advice President Roosevelt is thought to have received from his advisers on public relations. Dr. Edward P. Lilly believes that these advisers were opposed to central direction of an information program. The inevitable result was disunity in the voices that undertook to set forth our position to the world.

Another factor was the bureaucratic atmosphere of Washington. For example, in the early days of the Office of War Information a War Information Policy Committee was set up. It met only three times. At its first meeting either the directors or Assistant Secretaries of such agencies and departments as CIAA, the War Production Board, War, and Navy were present. But the Department of State sent only its public relations officer. At the second meeting, therefore, the War and Navy departments sent their opposite numbers to the State Department's representative. At the third meeting the representatives were all information chiefs, none of whom was empowered to make policy. It was never convened for a fourth meeting.

In addition, the information agencies were not made privy to our military plans. They had no idea, therefore, how a government agency should react to developments. We treated the fall of Guam as proof of our peaceful intentions and the aggression of Japan. But our defense of Wake we treated as proof of our fighting qualities with which we would defeat Germany and its allies.

We were operating in the early days of World War II on the general proposition that *truth*, in the long run, would defeat the enemy. The importance of a consistent program linked to our entire war effort had yet to be grasped and implemented. For example, at the time of the British offensive in Libya, which appeared to be succeeding, the Foreign Information Service was naturally inclined to make the most of it. British intelligence advised FIS not to do this, as the Royal Navy had not been able to keep Rommel from being reinforced, and the success was not expected to last. It was an FIS complaint that this kind of help was never given by their own counterpart members of the national team.

Furthermore, the tools of our psychological warfare remained in private hands. The Foreign Information Service would provide information to the private American broadcasting companies, but there was no assurance that they would use it or, if they used it, that they would stick to the text. Not until the time of the North African invasions was it decided that our official overseas broadcasts should be taken out of private hands and made the responsibility of government.

Lack of Confidence and the Security Problem

An additional complication was provided by the lack of confidence the military had in many of the information people, which, at the same time, was compounded by the information people, who apparently saw no need to consult with the military about what they proposed to say. Dr. Lilly, who has read extensively in this field, holds that in the days of the Foreign Information Service, which preceded the establishment of OWI, FIS would make an effort to find out what the major policy-formulating departments wanted done in the field of American propaganda, but FIS was not going to tell those departments what it intended to do or was doing. The natural reaction of the older departments necessarily was that, uncertain of the trustworthiness and security restrictions of FIS and having no knowledge of or control over FIS output, Army, Navy and State adopted the attitude that the least information, much less any outline of national plans and policies, given to FIS representatives, the less the possibility of official embarrassment.

How, then, did our information people find out what was afoot? Through the British. The British Political Warfare Executive (PWE), which was the counterpart to our propaganda organization, was an acknowledged member of the British national team, and as such it was given the information it needed about allied strategy in order to do its job. It shared this information with its American opposite numbers, both in London and Washington. In the time-honored witticism, this was "a hell of a way to run a railroad."

The situation got better, of course, as the war progressed. Both military and civilian officials were aware of the implications in this state of affairs and knew that it must be corrected. It was. But in what seems a rather typical American way. As they rubbed shoulders in their daily tasks, both sides grew to understand a little better their differing points of view and came to realize that good men held to each.

Nevertheless, although improvement took place, the issues that divided opinion on the aspects of intelligence and security have not yet been settled. The Joint Chiefs of Staff are on record that while they do not desire to formulate national policy, they take the position that it is their responsibility to control activities, even if these are civilian in nature, that might assist or conflict with military success. There are obvious points on both sides of the dispute. It is probable that no hard and fast decisions can be made prior to each instance that requires deciding. What seems necessary is effective machinery that will permit these differences of opinion to be resolved. This machinery is still not prepared and ready to function.

The Recurring Issue of Control

It was difficult, of course, to organize an effective program of information before determining a method for its control. This issue recurred throughout the war as various developments took place. Essentially it revolved around the question: Was the program to be under military supervision or civilian?

This same issue arose in World War I between the Army and the Creel Committee. In effect the settlement provided that the Creel Committee would be responsible for propaganda directed at allies, neutrals, and the enemy civilian population while the Army assumed responsibility for propaganda to the German soldier. This division of responsibility, while it might look sensible, in fact will not work, particularly with modern technological means for mass communication. A powerful transmitter in a military theater of operations, for example, can be heard far beyond the limits under the military operational control of the commander.

When we undertook psychological warfare operations in World War II, the problem arose again. The military took the position that such operations in a theater must be under the theater commander's control. But both OSS and OWI operators steadily refused to accept military decisions when the issues involved went beyond theater limits. And while the position of our war information activities was stabilized to a considerable extent by the executive order establishing OWI, our information functions were not centralized, and authority over them was not clearly distributed among military and civilian agencies. For example, to the end of the war CIAA had the field of information to the Americas virtually to itself. It was largely autonomous. Policy and operating powers, on the other hand, remained in the White House and the powerful State, War, and Navy departments.

Our overseas information operations therefore were required to adjust to the powers of these three departments with the White House in the background as a last resort. In effect, this meant that these departments could carry out information operations in their own fields while the Director of OWI, responsible directly to the President, had no authority over what was being said. The struggle to set up a coherent, effective system of control covering all United States information activities went on throughout the war and still was not settled at its end. Each military theater set up its own provisions for the control of psychological warfare autonomously while in nonmilitary theaters the Department of State insisted on a measure of control. Furthermore OWI and OSS added to the difficulty by struggling between themselves for control over each other's overseas propaganda activities. This struggle

rested on differences between those who believed that propaganda should form part of the program for subversive operations, and should consist of any action, true or false, responsible or irresponsible, which would effectively hamper the enemy at any point; and those who believed that propaganda should be a public, responsible, government operation to tell the truth about the war, about the United States and its allies, as means of describing democracy and freedom, our war aims, and our determination to win both war and the peace.[4]

The struggle between OWI and OSS was settled by an executive order that upheld OWI, but this order did not clear up the question of military or civilian control. It stipulated that OWI activities overseas in areas of military importance would have to be co-ordinated with the Army and Navy and be approved by the Joint Chiefs of Staff and that in military theaters the actual execution of programs was under the control of the theater commander.

Hereafter when OWI drew up its basic propaganda plans, the Army requested that a paragraph be included in each one that specifically claimed Joint Chiefs of Staff and theater commander control over any OWI activities in military theaters. On the basis of this paragraph the liaison officers to OWI from the Joint Chiefs of Staff could object to, or prohibit, the OWI Central Directive and its daily guidances referring to any items in a way with which the Joint Chiefs of Staff disagreed.

In effect, though from this time on OWI and civilian officials could draw up propaganda plans and directives free from military dominance, they could not implement them in military theaters unless the military agreed.

The Dilemma of Security

One of the most difficult and troublesome issues in the whole field of political communication is the problem of

[4] *Ibid.*, pp. 19-20.

what is called security. There are times when measures that involve national policy must be closely guarded and the steps contemplated known only to a few leaders in the government. Particularly is this the case in time of war. Nevertheless, if the men responsible for a program of political communication are not informed of these measures, it is almost axiomatic that our public information will be unable to plan effectively and, at times, will be shown up as futile.

This question is a vexed one. There are times when more will be gained by giving knowledge than by withholding it. Authority, judgment, discretion, and, most important of all, an understanding of the role information plays in affecting attitudes are the factors entering into all decisions between security and information. From the record, this has not been widely recognized in the history of our use of the instrument.

It seems still to be unrecognized and there are a number of indications that, if a situation similar to World War II arose again, security would still be the dominating concept as it has been in the past. The evidence, however, rather conclusively indicates that security ought to play a secondary role. In other words, the question to be asked is not: Does this violate security? But: Does this security violation gain more than it loses? There can be, of course, no hard and fast ruling that fits all instances. Each example requires decision on its merits. The point is, the decision-making officials must be sympathetic to, and knowledgeable about, political communication, its aims and capabilities.

During World War II, for example, OWI in Washington was not informed about the forthcoming landings in North Africa, even though its members were to participate. It first learned of the operation from a member of Britain's PWE. James P. Warburg, who was given the information overseas, tried to give Washington headquarters hints of the responsibility he faced in preparing psychological warfare plans for the landings, but security regulations kept him from being able to do more than make vague references.

It was not only the civilians who were kept in the dark. The Army's own propaganda branch in the military intelligence division was likewise left uninformed.

The liaison officers of the War and Navy departments with OWI served also as liaison officers from the Joint Chiefs of Staff. They were able to tell OWI of impending operations only when this was authorized by the Joint Security Control mechanism of JCS. And, once again, information planning had at times to proceed in the dark.

Reference has already been made to the dropping of the first atomic bomb on Hiroshima and Nagasaki and the fact that OWI was completely unprepared for this. The Darlan episode in North Africa is yet another example of a policy decision uncommunicated to the information specialists, which resulted in our propaganda being thrown back in our faces by the enemy.

Effective political information requires that our own plans and operations be known to at least the senior responsible official of an information operation. This, in turn, requires that such men have on their staffs individuals of sufficient caliber and discretion to be trusted with what at times will amount to the nation's security. This problem has not yet been honestly and adequately faced. It is reasonable to think that if it were, considering its importance to an effective operation, machinery could be devised that would be acceptable both to those responsible for the formulation of policy and to the departments responsible for military operations.

Knowledge of developments is closely linked to the problem of security. Without this knowledge it is impossible to explain and interpret intelligently. And there is always present the danger that what may be said because of ignorance will be proved false by events. Although this may have been recognized, it did not govern our propaganda operations in World War II or, for that matter, in World War I.

When the Armistice came in 1918, our military propagandists do not seem to have been informed about the details of the impending cease fire. At the time of the atomic bombing of Japan, they were caught completely unprepared.

Relations between the Theater and Washington

As World War II progressed, the problem of the theater and its psychological warfare operations came into the picture. The concept of the theater commander controlling such operations in his theater took deep root. It was unquestioned in the Korean conflict.

There were, in effect, some five psychological warfare operations going on in the different military theaters in World War II. In London there was the OWI. The operation under the aegis of the Supreme Headquarters Allied Expeditionary Force (SHAEF) was a combined one with OSS, OWI, and military personnel. The China-Burma-India (CBI) operation was essentially civilian in character. The Southwest Pacific Area (SWPA) was an OWI-military venture. In addition the Navy permitted an operation under the wing of the Commander-in-Chief, Pacific (CINCPAC), which was run by OWI.

In all these theater operations the organization varied. It is perhaps not too extreme to say that the only element common to all—with the possible exception of the London operation—was the concept that the theater commander had final authority.

In each of these theaters, however, OWI personnel played a leading part. And it was the OWI mission that concerned these men. This mission fell into four categories.

1. In enemy countries . . . to undermine and destroy the morale of the enemy population. . . .
2. In enemy-occupied countries . . . to keep alive hope of liberation and stimulate resistance to the enemy forces of occupation. . . .
3. In neutral countries . . . to win the moral support of the neutral population. . . .
4. In allied countries . . . to counter all enemy propaganda aimed at dividing the United Nations, to promote the morale of our allies . . . and to foster a better understanding of the United States.[5]

It is conceivable that a theater commander might well find a propaganda operation under his jurisdiction concerned with all four categories. A mission of this scope, self-evidently, should involve major national policy and decision at the highest political level. The highest political level can not very well step down to the theater. The writ, obviously, should run the other way. Particularly is this the case when the kind of warfare we may have to wage in any future war is taken into account.

It is in the theaters, particularly, that we observe the unlovely fruits springing from a lack of psychological warfare co-ordination. OSS had been given the field of black propaganda. OWI had white. Both agencies were under the theater commander's control, and he, in the final analysis, was the authority—not Washington. Whether or not either, or both, carried on operations in his theater was left to his discretion. OSS, for example, was not permitted access to the SWPA theater while, in this same theater, OWI, although having a team present, found that its weekly directive from Washington was not acceptable. Its team, therefore, was forced to operate under alien instructions.

Two main points in the history of political communication

[5] *Ibid.*, p. 40.

may be noted as we study the movement of psychological warfare into the military theaters. The first is the dominance of the theater commander. The second is the refusal of the agencies to accept his supremacy in matters going beyond his theater. This issue has not yet been settled.

In fact none of the issues discussed so far was settled by the time the war came to an end, and, in spite of all that was accomplished—and it was not inconsiderable—our psychological warfare operations ended the war never having been properly organized or co-ordinated as a national instrument smoothly functioning as an element of the nation's accepted departments in the conduct of international affairs.

The Korean War

For a time, after World War II was over, we once again neglected the lessons that had been so hard to learn. The men who had gained experience with this weapon were largely scattered, going back into more peaceful pursuits. Nothing at first was done to glean this experience and winnow it for analysis, in case, once again, it became necessary to call on it.

Three years after the war ended, however, the Army was evincing definite interest, although in the Joint War Plans Branch of G-3, the Pentagon, only one full-time and one half-time officer were attached for psychological warfare purposes. Nevertheless, with the organization of the Operations Research Office of The Johns Hopkins University in 1948, one of the first projects the Army asked it to undertake was an attempt to determine "the maximally effective weapons, instruments and techniques that may be employed by ground forces in the conduct of psychological warfare operations."[6]

6 From the Mission given to ORO by the Army.

To their considerable surprise the staff analysts of the Operations Research Office found that many of the assumptions they had made at the outset of the project were not supported in fact. It was assumed, for instance, that from the wealth of data and experience accumulated in World War II the information on which research planning had to be based was in existence. This assumption proved false.

> Basic doctrine on PW [Psychological Warfare] was assumed to exist in the general system of Government agencies concerned with the problem and of academic and other experts on the problem. The fact that the whole body of knowledge on PW is to this date merely an aggregate of empirical data, historical observations, theoretical controversy, *a priori* theses, was not formally established by ORO for its own purposes until recently. In short, it was assumed at first that there were some masters of the problem, authorities who had a systematic understanding of its full scope. The growing realization that no transmissible reliable fundamental true body of knowledge exists on psychological warfare demanded, and still demands, basic research on the problem so that valid operational research can be built up. The lack of a general doctrine on PW is now recognized.[7]

This was written in August of 1950, five years after World War II had ended and two months after the Korean war had begun. In this war, peripheral as it was, unexpected and unprepared for, psychological warfare once again was called on to see what it could do. And, for the first time during a military operation, a team of operations analysts was requested by the Army to assess the psychological warfare operations and determine how its military application might most effectively be employed.

The experience in Korea unfortunately paralleled far too closely the experience in World War II. Once again we were unprepared. Once again security hampered the operations.

[7] Operations Research Office, The Johns Hopkins University, internal Memorandum.

Once again intelligence was faulty. Once again the organization was imperfect. The lessons learned in World War II largely had to be learned all over again. They were not in collected, transmissible form as a guide for the Korean operation. And few of the experienced operators of World War II were called on for their services.

The United States Information Service had a large information operation under way in Korea when the parallel was crossed. Hardly any of these men were drawn into the Army's own operation. In World War II, psychological warfare had been an amalgam of soldiers and civilians. In Korea it was soldiers. It was a military affair from start to finish. Its success was limited. The files of the Operations Research Office and its studies published for the government amply document this statement.

Eighteen months after the outbreak of hostilities, the Theater Staff Organization for psychological warfare still was not fully exploiting the potentialities of the weapon on the evidence of an analysis of the propaganda that was being put out. Furthermore, the Psychological Warfare Section of the Far East Command was virtually autonomous despite the fact that the Department of the Army, soon after the war started, had created a special staff section for psychological warfare responsible to the Chief of Staff of the Army.

Beyond this the Far Eastern operation was proceeding independently of political authority on many of its most urgent problems of propaganda policy. Intelligence still was not being tailored to psychological warfare needs. The strategic radio operation felt the lack of adequate preparatory measures in terms of organization and training. Only in the area of tactical psychological warfare—the combat operation in or close to the front line—was improvement noted. In this area the Army scored distinct advances over its procedures in World War II.

It seems clear, in looking back over the history of psychological warfare in both world wars and the Korean conflict, that we are not entitled to consider that it was at any time a planned operation fully integrated with national policy. Partly this must be because, on certain issues, no specific national policy existed. Partly it must be because we were not practiced in crystallizing foreign policy. When psychological warfare is faced with rigid requirements—for example, that the enemy must surrender unconditionally—when it must conduct its operations without knowing what postwar policy will be followed, when, as in the Korean war, it must act when the logic of the war itself has not been fully specified, then planning, organization, and terms of reference obviously must suffer.

From the three wars in which we have used propaganda as an instrument of national policy, there is now sufficient evidence to support the conclusion that our planning has not been brilliant, our organization has not been integrated, and jurisdiction has not been faced and decided in terms of national policy for a democratic government. In other words, the basic requirements for a well-knit, co-ordinated psychological warfare operation have not been logically thought out and then fastened down with appropriate directives.

V

An Instrument of the Cold War

The roots from which the concept of political communication has sprung are in war. The term in general use for the instrument is *psychological warfare.* Even while calling it this, however, we were in fact trying to use it in contexts where the term was misleading. For example, one aspect of OWI's work was to promote the morale of our allies. This is hardly psychological warfare, unless a definition is considerably stretched. Nevertheless the name has stuck, and in recent years all our activities in this field have been lumped together, and, loosely, we have named the bag that holds them psychological warfare.

However, while this term is generally accepted, in practice it is clear that the concept behind it has been growing and maturing. From considering this an instrument concerned essentially with what are technically called symbols, we have begun to see that, in fact, it encompasses much more—all that may be construed as having to do with the psychological posture of a given situation.

With the end of World War II, OWI as such was disbanded. But some of the work it had been doing was taken over by the Department of State. In effect it reverted, as Lilly says, to being an information program, and as such it was hardly thought of as a weapon, but rather as an adjunct of government with some attributes that might be of advantage in the business of diplomacy.

During the last ten years it has been developing until once again it is being thought of as a weapon, and the manner of its use is that of a weapon. But this has not been a planned and orderly development. The logic of political realities has forced its maturing. It is not yet, however, by any means the powerful instrument its history and capabilities indicate it could become.

Along with this development of an information service in the direction of becoming an instrument of state, the concept of what this book calls political communication also was developing. In other words, two separate activities have been going forward within the government, and both give indications of converging. One of these is the growth of the information service, and its chronology probably will be of value.

The immediate successor to the Office of War Information was the Interim International Information Service, Department of State. Within a few months this service became the Office of International Information and Cultural Affairs (OIC).

In the fall of 1947 OIC became the Office of International Information and Educational Exchange (OIE). Early in 1948 the passage of Public Law 402, variously known as the Information or Smith-Mundt Act, established an information service for the United States. Soon after the passage of this act OIE was transformed into the Office of International Information (OII) and the Office of Educational Exchange (OEX).

By 1952 these had been recombined into the International Information Administration (IIA). In 1953 IIA was reorganized as the United States Information Agency (USIA) and given separate standing apart from the Department of

State, under whose aegis, since the abolition of OWI, it had functioned.

Concurrently with these outward and visible signs of activity in the field of international information proceeded an inward and invisible succession of conferences and meetings addressed to the following basic points:

1. Did a need exist in time of peace for a psychological warfare plan based on existing operations that would serve as a nucleus around which a full-scale operation would develop in time of war?
2. Was it necessary in peacetime to establish an organization to co-ordinate and advance psychological warfare planning in terms of changing national policies and objectives?
3. Should psychological warfare be under civilian or military jurisdiction in theaters of war?
4. Should white and black operations be combined under one agency or should the Central Intelligence Agency supervise black operations, if and when required by war?
5. To what extent should the United States co-ordinate psychological warfare planning with that of allies, particularly the British; and what role should NATO play in such planning?[1]

The work of the committees that discussed these and allied questions, in the years following World War II, shows us the direction in which our thinking in this field has been developing. Originally this work began at the direction of the State-War-Navy Coordinating Committee (SWNCC). It continued under SANACC (State-Army-Navy-Air Force Coordinating Committee.) Later it was directed by the Interdepartmental Foreign Information Organization (IFIO),

[1] Point 4 requires separate treatment and is dealt with in a later chapter. Point 5 can be meaningfully discussed only in terms of the actual deliberations, which is not yet feasible.

which was the policy-making board for the Interdepartmental Foreign Information Staff (IFIS). Later still came the Psychological Operations Coordinating Board (PCB) and the Psychological Operations Coordinating Committee (POC).

The relationship of these various committees to the work done is omitted here as of less interest than the work itself. It is necessary to remember only that the men on these committees and their staffs represented the Department of State, the Department of Defense, the Central Intelligence Agency, and what is now USIA. At various times there were additions, but the core was formed by men representing the departments given.

We can properly understand and assess the developments in this field after World War II only if the pattern of governmental procedure in our society is comprehended. We were confronted, in effect, with a new instrument that few as yet fully understood. We had moved, by virtue of our delivery of atomic weapons on Japan, into the atomic age. We had emerged from the war as one of the two most powerful nations in the world. The necessity for leading and the responsibilities of international leadership were being thrust upon our attention. In our beginnings, as we confronted a new age, our methods of approach followed customary patterns. And customary patterns make it difficult to rethink traditional concepts and re-examine lines of authority challenged by revolutionary advances in knowledge and technology.

One result of this was that the membership of the committees concerned with the problem of political communication faced it in terms of their services and departments and in terms of what their experience led them to consider the logic controlling the decisions they were called on to recommend. The instrument clearly was thought of as capable of separation into independent component parts. In fact, little recognition seems to have been given at first to the nature of the relationship between operations in war and operations in peace.

The First Phase

It was in terms of war that the problem was approached at first. And, in this context, appropriately enough, it was the office of the Secretary of War that initiated action. Early in 1946 it was suggested to the office of the Secretary of the Navy that a civilian-military group of qualified individuals tackle the problem of planning how psychological warfare should be organized and conducted in the event war again broke out. A year later a subcommittee took up the question under the direction of SWNCC. During this year a considerable amount of thought and change in emphasis obviously took place. Originally the problem was viewed as one of concern only to the Army and the Navy. The clear intent was an organization to go into action in any future war. And the aim was to prevent a repetition of the organizational errors of World War II. But equally clear is the indication that the activities of the information service under the Department of State were not thought of at first as an irrefragable component of the question.

By the spring of 1947 the SWNCC subcommittee appointed to make the necessary plans had enlarged its terms of reference. It was still in terms of war that it approached its task, but it was also clearly recognized, and written into the appropriate directives governing its work, that no plan could be drawn up without taking into account the role of the Department of State and the work of the fledgling Central Intelligence Agency.

Equally clear were the individual assessments of psychological warfare in World War II by the various departments and agencies of government concerned. For example, it was generally agreed that the weapon had come to stay. It was also agreed that the Department of State had primary interest in the determination of policy for the activity when United States foreign policy was involved. The Defense Department staked out a claim for policy affecting our national security and military operations. Everyone piously agreed that everybody in government must help with the activity and not leave the baby out on the doorstep, as had been the case in World War II. And the Defense Department staked out a further claim by taking the position that the implementation of psychological warfare in theaters of war would be the responsibility solely of the commanders concerned.

This shows quite clearly that the military role of the weapon was given preponderant weight and that political authority virtually was relegated to stating the foreign policy objectives of any psychological operation in war. This position was not acceptable to those members of the subcommittee with OWI experience.

Quite evidently an attempt was being made to resolve the terms of reference of psychological warfare about which there had been so much friction in World War II. The military position sought to prevent any further situations in which the theater commander's authority was challenged. Furthermore, the military did not want directives coming through its channels calling for propaganda activities it questioned in the name of security. This is entirely understandable from the military viewpoint. Not so understandable is the position of the Department of State, which was lofty and olympian. Its concern was with foreign policy. Psychological warfare objectives must not conflict with this policy. Beyond this, even though the disbanded OWI had been re-created and put under State Department direction, the attributes of the instrument, clearly apparent by virtue of its work in World War II, were of little concern. The

Department of State was not particularly interested in talking to massed audiences. Its history and tradition were those of the diplomatic conversation with the right, select company—the gesture that said nothing and made no commitment but was indicative. It had so far failed to grasp the implications of the instrument as a legitimate and honorable means for informing the minds of men and mobilizing them as a force on the burning issues the atomic age was almost daily bringing to the world's attention.

What was developing—and perfectly naturally so—was a struggle between two powerful departments of government, State and Defense, over the question of where control of the instrument should lie. What is of importance to note is that at this stage few grasped the logic of the instrument and the fact that its nature required it to be directed in terms of our over-all national objectives, and not in terms merely of diplomatic or military aspects of policy.

In this struggle there were a few who sensed—perhaps, even, had clearly grasped—this attribute of what had come to be called psychological warfare. Their voices were raised repeatedly in attempts to gain a broader approach to the problems the working committee faced. But they were caught between the upper and the nether millstones of the two most powerful departments in the government with not only important decisions of organization, planning, and procedure at stake, but also such factors as prestige and traditionally different casts of mind affecting viewpoints, according to how one viewed the diplomatic and military arts and their roles in national affairs.

By the end of 1947 the members of the subcommittee charged with the responsibilty of recommending an organization and plan for psychological warfare had begun to see that they could not proceed without taking into account our peacetime activities. It was observed that we were not using co-ordinated psychological measures to counter the Soviet propaganda campaign and that our program of foreign economic aid was not being explained and interpreted so that it might be expected to achieve maximum effect. There is, in this recognition, the beginnings of the concept called in this book political communication. But it was only a beginning, and the emphasis and direction still were in the direction of a weapon for use in the event of war.

However, it seems accurate to say that after a year's concern with the problem, the differences of opinion still were preventing the formulation of a plan on which there was general agreement, and in the spring of 1948 the Armed Forces were pressing for an independent agency to handle psychological warfare, and they were supported in this by the Joint Chiefs of Staff. Within the Department of State this was strongly opposed on the ground that except in time of war the activity was essentially political in nature, and responsibility therefore had to rest with the Secretary of State. Furthermore, the position was taken that at virtually the highest level the Secretary had been given this responsibility.

The Department of Defense nevertheless pressed its position on the ground that the national security would be adversely affected if a going organization were not set up prior to war. But a compromise was effected on the question of the actual creation of a working agency. It was agreed that an organization should be established, but that its peacetime scope should be limited to that of preparing plans and working out the machinery for co-ordination to go into effect if war occurred.

This was an uneasy compromise and within the Department of State was still opposed. The Department of State took the position that it should be responsible for any planning during peace, but suggested that there should be military participation in the planning. Those familiar with the bureaucratic machinery and procedures of Washington will find it difficult not to feel that this question of where the planning organization should have its home was one of the root difficulties. Obviously the agency within which it was established would be in an advantageous position to influence the directions and decisions taken.

Three years after Secretary Patterson had initiated action—that is, in the spring of 1949—the highest arm of the executive made known its stand. It established within the Department of State a small organization to plan for psychological warfare in the event of hostilities. But this, in itself, was a further compromise rather than a resolution of conflicting conceptions. What the military establishment desired was an independent agency that could make plans for psychological warfare that would be implemented if war broke out. In peacetime this agency would have played a stand-by role, its terms of reference putting it into action only if we fought. The civilian viewpoint called for an agency that could not only make plans, but could also integrate those plans with the on-going information service activities operating under the over-all direction of the Department of State.

The compromise decreed accepted the military contention that a plan for use in war was necessary, but gave to the Department of State the task of supervising its drafting. The question therefore of a truly national program of political communication, integrated with the activities of government, was still being begged. There was also apparent a fundamental difference of opinion on the nature of the instrument itself. It is clear that at the end of the first three years after World War II there was a school of thought within the staff working on the problem that held that the activity was an integral instrument for furthering national policy, equally applicable in peace or war although obviously with different emphases according to circumstances. The second school held that the instrument was applicable only in time of war.

The first school may be said, therefore, to have been reaching toward the concept of political communication. The second was concerned only with that branch of political communication that may indeed properly be called psychological warfare. The second school did not consider that our peacetime activities in this field bore any real relation to the activity in war. The two activities were distinct and separable, and any planning and organization should proceed on this basis.

It is interesting to note that the staff members who held to this second viewpoint were representative of the military establishment and the Department of State. The smaller school, which considered that the activity was one, whether in peace or war, was formed of those civilian representatives who had had actual experience with psychological warfare as an operational activity during World War II.

The position of the State Department in viewing the activity as essentially warlike of course calls in question the logic of its being so insistent that the planning and organization of psychological warfare should be under its direction. It must be kept in mind that the on-going information service was at this time within the Department of State. It was through the Department of State, therefore, that the representatives of the information service made known their thinking. In reality, therefore, the State Department position on planning and organization really represented the concept of those with experience with psychological warfare in operation and familiar with its history and the record of our wartime experiences, obstacles, and frustrations. The State Department was the mouthpiece for information service representatives—the only mouthpiece they had, since they were part of the department. On the question of the boundary determination—the question of whether this was an instrument for peace as well as a weapon in war—the presence of the State Department alongside the military establishment in truth represented the department's own thinking as distinct from the thinking of the information service.

The Second Phase

The peculiar difficulties an organization conducting political communication faces, and the strength and tenacity with which differing opinions were held, may, in some measure, surely be gauged by observing that after three years—spring, 1946, to spring, 1949—nothing in effect had really been accomplished as far as an actual plan for the conduct of psychological warfare in the event of hostilities was concerned.

The second phase lasted some six months—from the spring to the fall of 1949. The first step in this phase was the formation of a small group of representatives from the Department of State and the National Military Establishment working under an Assistant Secretary of State. Their task was to plan and make preparations for the co-ordinated conduct of psychological warfare in the event of war. At least some of these representatives continued from previous committees.

In the six months occupied by this phase, the deliberations centered on two problems both involving jurisdiction. The first was the issue of civilian or military control of the activity. The second was the issue of intelligence support.

The position of the civilian representatives may be given in one sentence, which was repeatedly expressed in one form or another: It is imperative that planning now for overt propaganda should ensure civilian control in wartime, particularly foreign policy direction.

The firmness of this stand is matched by that of the Defense agencies, which, equally firmly, repeatedly put forward organizational positions that, in the event of war, assured them of *de facto* control of psychological warfare in all areas in which military interest and action were predominant. The military position was adhered to with such disciplined insistence that the civilian representatives, acting apparently on the theory that nothing could be accomplished if both sides refused to budge, concluded that the principle of military control in theaters of war should be accepted, provided military commanders were required to conduct all psychological warfare operations within the framework of approved policy directives.

This concession, however, was not satisfactory. The military establishment, undoubtedly remembering its difficulties in World War II, wanted to be sure it could fight and maneuver with the utmost freedom in the event the national survival was threatened, and it pressed for a military veto power over any peacetime plans to be implemented during war and actual control of military operations in this field in the event of war itself.

The point must be reiterated and re-emphasized: neither side was really clear as to what the activity it called psychological warfare was all about. The military position was no intransigent stand springing from a command complex. It was based on hard experience learned in World War II when theater commanders found that their wishes and instructions to psychological warfare personnel were not complied with if, in the opinion of the psychological warriors, the issues transcended theater operations.

By the same token, the stiffness of the civilian stand rested on precisely this same experience, and it was on the ground that issues would in fact arise transcending the theater and requiring political judgment to resolve, that the civilian representatives of the working group stuck to their guns.

The significant point is that neither side turned to the one question that conceivably might have helped to provide a clear solution, namely, the nature of the activity itself and the premises on which, in a free society, it must be undertaken. This factor grows increasingly clear as one observes the hard and earnest attempt made by honest, intelligent, and dedicated men to forge an instrument that would really serve the interests of the United States.

The issue of intelligence was the second stumbling block. All those who have had any experience with the instrument of political communication regard intelligence as one of the essentials to successful operation. Once again it was a jurisdictional dispute. CIA was a fledgling agency and jealous of its status. It also carried a nucleus of old OSS hands. OSS and OWI were at loggerheads on a number of issues in World War II, notably on the question of the relationship of white operations to black.

In this second phase of the development of political communication as an instrument of state, CIA did not question the need for intelligence. It did question who was to provide the organization for that intelligence and decide what intelligence the operation ought to have. The answer, of course, is self-evident. CIA picked itself for the job and refused to permit an organization for psychological warfare to attempt to decide what it ought to have. One further complicating factor was the traditional nervousness of any new agency in bureaucratic government operating with older and established departments who will not view its appearance with affection. CIA intended to establish its position, and this came first. The issue remained unresolved in spite of protracted discussion, many meetings, and much putting of pen to paper.

By the fall of 1949 there had been much talk, much fencing, jockeying, and search for accommodation of conflicting viewpoints, but there was still no organization for psychological warfare. It is difficult to escape the conclusion that what was, in reality, at issue was never brought out into the open or expressed. Simply stated it was this: Who will have the power to run psychological warfare, *and what is it?*

After more than three years of attention to the problem of psychological warfare by the Department of State and the military establishment, there was little to show for it. It is true that a group had been appointed with the mission of drawing up a psychological warfare plan. No plan as yet had been drawn up and approved. Certain methods of procedure had been canvassed; for example, the State Department would determine the policy, and in the theaters military command would execute it. What stands out clearly is that nowhere do we find any discussion of the role political communication should play in the national policies of a free society, either in war or peace. The basic premises, in other words, were implicit, but they had not been analyzed. On the principle of first things first it must seem hardly surprising that a number of agencies found difficulty in arranging a meeting of minds when the role of the instrument under consideration had not been defined, nor its place among the established agencies constructed for the security and welfare of the nation.

What the record unmistakably reveals is time and energy and ability being concentrated on an attempt to formulate a plan for the undertaking of psychological warfare without the various components of the committees concerned together being in agreement as to what, colloquially stated, political communication was all about.

The Third Phase

The third phase may be said to begin with the organization of the Interdepartmental Foreign Information Staff. Its members were drawn from the departments of State, Army,

Navy, Air Force, National Security Resources Board, and the Central Intelligence Agency. Later a representative from the Economic Cooperation Administration was included. Its work was directed by the Assistant Secretary of State for Public Affairs. The pertinent documents relating to this phase appear to indicate that the allegiance of these members was not to IFIS, but to the parent agencies from which they came. The staff was divided into two standing committees—one for current programs and the other for advance planning.

By this time the concept of political communication was becoming clearly apparent, although the activity was still being called psychological warfare. The lengthy discussions preceding this stage, the trend of developments in the world, the realities of political life, all combined to force further attempts at organization and planning in the direction of viewing "psychological warfare" as a continuing operation, to be carried on in both peace and war. The task had become one of determining how best this might be achieved.

During this stage a psychological warfare plan at last was drawn up. It may not be considered final. It had not resolved the strongly held opposing viewpoints of the various departments of government. It would have been put into operation if necessary. It would not have resulted in an effective operation. But it was a step forward. Some machinery, faulty as it was, had reached the stage where the method of its working could be set down on paper. A great deal remained to be done. Nor should it be thought that drafting various plans and enabling moves to supplement them were simple tasks. Bureaucratic government, in which committee representatives owe their allegiance to the departments from which they are on loan, does not work this way.

The simple combination, for example, of two previous

directives authorizing a plan for psychological warfare into one later and more complete form required six months and the drafting of some twenty copies, which were circulated among the members of the staff, before a final copy reached the stage of agreement.

One of the vexed issues on which divided positions were taken concerned an interpretation of CIA's co-ordinating function in intelligence. There were those who felt that the new agency was attempting to use the staff to settle the intelligence responsibilities of CIA with respect to all government departments and not just its responsibility for providing the necessary knowledge to those engaged in political communication. This issue was the crux of conference after conference among the members of the staff. By the fifty-ninth conference, held in May of 1950, it was apparently recognized that for the time being CIA would not budge from a position in which it was a minority of one. And in order to proceed with other aspects of the organization and relationship of an agency for political communication to the other departments of government, the issue was shelved.

Another problem that caused much heart-searching and protracted discussion revolved around the still unsettled question of jurisdiction and procedure. It was proposed, for instance, that there should be established within the State Department an organization authorized to make recommendations to the Secretary of State on aspects of the activity that involved interdepartmental action. It was argued that the proposal strike out the authorization specifically naming the Secretary of State as the recipient of recommendations on the ground that this would prevent recommendations to other departments of government. The Department of State objected, arguing that if recommendations were made to other agencies, they should still go through the Secretary.

It was points like this that consumed much of the time and attention of the staff in their attempts to draw up and make explicit the lines of authority within which a national program of political communication should function.

The old issue of the responsibilities of the Department of Defense and the Department of State in peace and war also had to be restated and discussed and argued anew. And, as in the past, there was little if any visible move in the direction of a meeting of minds among the protagonists of already entrenched positions.

Much discussion also concerned an organization to conduct political communication. Four possible alternatives were considered. One was an organization answerable to the President. Another was a separate agency. A third was an organization answerable to the National Security Council. The fourth suggested that the policy-making board for IFIS, known as the Interdepartmental Foreign Information Office (IFIO), be given a new director and take over operations within the Department of State.

In fact, none of these alternatives came into effect. Instead, after deliberation, President Truman authorized a Psychological Strategy Board (PSB). It would appear that this board was a direct outgrowth of a growing recognition in government that neither the Department of Defense nor the Department of State seemed able to resolve their own difficulties and agree on an organization acceptable to both. There was mounting dissatisfaction with the way developments had been progressing, and a strong move was afoot to pull the whole area covered by political communication out from under the wings of either the Department of State or the Department of Defense. In fact, this is what took place, and in June of 1951 the Psychological Strategy Board appeared on the scene.

It was not accepted without misgivings by the Department of State, which felt that the direction and control of our psychological activities ought to be under its own eye. Strenuous efforts were made to forestall the establishment of the board on the grounds that policy control operations would be taken from the Department of State and that acceptance of the President's directive might well lead to the loss of policy and operations in the foreign information field.

The board was created nevertheless, and it was a compromise attempt that was unsuccessful. It consisted of the Deputy Secretary of the Department of Defense, the director of the Central Intelligence Agency, and the Under Secretary of State. The board was to appoint its own chairman. From the outset the information representatives feared that the combined weight of the Defense Department and CIA might lead to the political factors inherent in a national program of information being relegated to a secondary position. In fact, it transpired that the board was unable to achieve a more unified approach to the problems of political communication than had been the case before it was created.

It struggled along for two years, and its lack of success was accurately described in the words of the Senate committee that investigated our overseas information programs.

> The Psychological Strategy Board was established to coordinate the psychological policies of the United States. Much of its activities are necessarily of a highly classified nature, but data has been available to the committee which suggests that the Board is not effectively discharging its function. The Executive order (June 20, 1951) creating the Psychological Strategy Board is loosely drawn. It lodges various responsibilities in the Board but limits its authority to "guidance" and "reporting." Lewis K. Gough, commander of the American Legion, testified that he has been advised that—"not a single major recommendation of the Board has been executed by the departments or agencies to which it reports." This report was not challenged by representatives of the executive branch. Observations of members of the committee abroad and other evidence moreover suggest that the component agencies of the Board (CIA, Defense, State) are continu-

ing to go their separate ways in matters of psychological policy.[2]

The Senate committee made public its findings on our information programs in the summer of 1953. Prior to this, however, the President took action. It was clear that the Psychological Strategy Board was not fullfilling expectations. In January of 1953 the President appointed a committee to study the field of political communication and report to him by summer. Its members were important individuals in business and government. Its chairman was William H. Jackson. Officially titled The President's Committee on International Information Activities, it was known as the Jackson Committee. The report it made is one of the more important and significant documents issued in the field. The concept of political communication is a dominating note, although this term was not used. It is impossible to read the report, however, without being aware of the emphasis the committee placed on the necessity for regarding a national propaganda operation as integral with all the actions of government. The direction of its thinking is clearly given in a forceful statement, which, in various forms, constantly appears throughout its pages.

> There is no "strategic concept for psychological operations" separate and distinct from a strategic concept for gaining national aims without war. . . . We find that the "psychological" aspect of policy is not separable from policy, but is inherent in every diplomatic, economic or military action. There is a "psychological" implication in every act but this does not have life apart from the act.

In accordance with recommendations in the Jackson Committee report, the President abolished the Psychological Strategy Board in September of 1953 and replaced it with

2 Report of the Committee on Foreign Relations, Overseas Information Programs of the United States (Hickenlooper Report), Senate Report No. 406, Eighty-third Congress, First Session (1953), p. 6.

the Operations Coordinating Board. The members of the new board were the Under Secretary of State, the Deputy Secretary of Defense, the director of the Foreign Operations Administration, the director of Central Intelligence, and the Special Assistant to the President for psychological warfare. In addition, the Special Assistant to the President for National Security Affairs and the director of the United States Information Agency were authorized to attend its meetings. In the spring of 1955 the director of USIA was formally appointed a member of the board.

OCB, as it came to be known, is still in existence. Insofar as the subject of this book is concerned, its main task has been the establishment of an effective organization in the field of political communication. It, too, has failed to bring together the opposing viewpoints held by representatives of the Department of State, the Department of Defense, and the Central Intelligence Agency.

It will be remembered that one of the objections to PSB was that it had no authority. It was limited to guidance. OCB, similarly, lacks authority to formulate a single, unified course of action when opinions differ. Its deliberations reflect and closely parallel the experience of the working committees that, in the early years after World War II, took up the subject. The members of OCB are first and foremost departmental representatives, only secondarily members of a board designed to co-ordinate and strengthen the voice with which government speaks. Thus, in practice, it has been possible for any member of the board to veto any proposed action in the field of political communication because its chairman has no authority to proceed on majority decisions. The board must agree as a unit. And if any member objects in the name of his department, that department is able to prevent the board from reaching anything but a compromise decision.[3]

[3] In February of 1957 the President made OCB a part of the National

As matters stand today we are committed to an international information program. And there is no doubt that, in the event of war, we shall be committed to a psychological warfare program. There is also a growing realization that our activities in this whole field, whether in peace or war, bear a close relationship to each other and are not separable. Nevertheless, the fact is we have not yet fashioned this instrument, either organizationally or conceptually, into the kind of instrument its history and capabilities indicate are necessary for effectiveness. Russia is still ahead of us in every respect but that of the first basic premise governing the field. Demonstrably the Soviet propaganda campaign is not based on the truth. Equally demonstrably the United States is attempting to give this premise force. In every other respect the Soviet organization is ahead of us.

It is well organized. It spends more money. It employs more people. It has trained those people. Its policy and its propaganda are integrated and reciprocal. It is time, if we would compete intelligently and with maximum effect, for us to concentrate on a workable, democratic series of decisions that will permit the instrument to speak with conviction and emphasis in support of the ideals, aims, and aspirations of free men.

It is difficult to summarize history of so kaleidoscopic a nature as that which records the vicissitudes of the development of the concept of political communication in the years following the close of World War II. Its predominant features are contradiction, conflict, and wrangling. On the record there is little to show by way of positive accomplishment. After ten years of attention to the problem, this question is a fair one: What have we got to show for the time

Security Council. The intent is clearly to strengthen the relationship between the formulation of policy and its implementation. It is too early yet to judge whether, in fact, this is the remedy. The organizational requirements and needs are discussed in a later chapter.

and effort and money spent in attempting to develop a national operation? The answer must be that we have virtually nothing. We have made little headway, and we are not much further along than we were at the beginning.[4]

If it is granted that political communication is in fact capable of becoming a powerful instrument of national policy, then, surely, in view of the revolutionary state of the world situation and the responsibility we have assumed for the leadership of free men, the premise follows that we cannot afford this sort of irrationality. To continue as preposterously as we have begun is to risk our security. It may be stated as simply and as grimly as that.

It seems clear from the history of political communication that what we have done so far is to put the cart before the horse. We have conducted skirmishes all along the line to determine responsibilities, administrative procedures, and jurisdiction without first planning a determined attack to reach an agreement on what it is that we are trying to do. Self-evidently this is marked for failure. And, on the record so far, we have failed.

The first question to which an answer should have been found is: What is political communication all about? Once this is clear, a rational, intelligent people, among whom are

[4] It should be obvious that no reflection is intended *per se* on the development and work of the United States Information Agency. This agency has indeed moved forward in the past ten years. But it is not yet an integral part of government. It is not, for example, being generally consulted on, and informed about, the more vital aspects of our national policy. We do not yet possess an organization so integrated with government that it is able to speak authoritatively in the name of the United States on the issues that the administration of the day believes significant and important as these are determined by the political, military, and economic actions it takes. The discussion here is in terms of such a national organization, understood and supported by all departments of government and so explained and interpreted to our own people as to have the support of public opinion. This kind of national operation in the field of political communication we do not have as yet.

many demonstrably dedicated to their country's best interests, could be expected to provide workable solutions to the problems of control and jurisdiction.

One patent fact seems to emerge clearly from the historical record. It is that, as a national instrument, this weapon must support all the branches of government responsible for policy and operations. It can never be the creature of a single department. It must lend its emphasis, at any given time, to whichever department of government has primary responsibility for attaining national objectives—economic, military, or diplomatic. This fact, it is clear, has not been recognized. The record shows that much effort has been expended on trying to decide who runs it in peace and who runs it in war. The fact is it is the same weapon in either instance; only its emphasis and the manner of its use vary. The premises that govern its application remain constant, and these must give the clue to its direction and method of control.

It is a fair conclusion that, if a fraction of the time that has been spent in arguing about who did what and with which and to whom had been spent in determining the premises and clarifying the role of political communication, many of the problems of administration would by now have fallen quietly and naturally into place.

VI The Tools and Requirements of the Instrument

One of the major difficulties we face in attempting to order the use of political communication for effective results is the absence of exact knowledge. In this area we are by no means able, in our arguments, to take the doubting Thomases into a laboratory and prove points at issue. On the other hand, this is not always either desirable or necessary. "Experimentation is one of the most useful techniques of science, serving to isolate the various causes of events used to verify assumptions derived from observation. Nevertheless it is not essential. There are no experiments in astronomy nor was the Darwinian Theory of evolution based on experimentation."[1] What we can, and must, do is seek "the careful and scrupulous use of the human mind."[2]

Political communication is both an art and a science, although at this stage of our knowledge it is perhaps more accurate to say that the various sorts of knowledge that are most useful to it are closer to being prescientific in their development. If we wish to get the maximum possible results from it, therefore, we probably have to "make haste

1 William Esslinger, *Politics and Science* (New York: Philosophical Library, Inc., 1955), p. 21.

2 *Ibid.*, p. 22.

slowly," which means that the operation itself must continually be subjected to critical analysis, both by those engaged in it and those who are interested in the theories of its use.

It is in the doing that political communication is best explained, for we seem, as yet, to be only on the threshold of scientific knowledge in this field. In the doing there is a place reserved for art; the art of the driving man of action, which, nevertheless, recognizes "the existence of systematic attempts of inquiry, which, when they are brought to bear on a range of facts, enable us to understand them better and to control them more intelligently, less haphazardly and with less routine."[3]

In terms of action, then, and with "the systematic application of reason as best we are able to do,"[4] what support can the practitioner call on?

In the first place, there is little of what is called in military circles "doctrine" on which he can lean with assurance. It has been evident for some time to students in this field that there is a need for a collection of facts bearing on past operations, which can be studied quantitatively to determine whether there are patterns in political communication that recur and may be governed by definite maxims. For example, in a minor key, does one *always* argue with the enemy in psychological warfare? Or *never* argue? Or blend the extremes? In World War II the "doctrine" was, never argue. In the Korean war it was, sometimes. It was one of Hitler's principles to propagandize on a few critical points with constant repetition. Hitler was a skillful propagandist. But it does not follow that this piece of "doctrine" is applicable to an operation conducted by a free society.

During the Korean war one analyst, contrasting our own psychological warfare with that of the Soviets, observed

[3] John Dewey, quoted in William Esslinger, *ibid.*, p. 22.
[4] *Ibid.*, pp. 80-81.

that the Communists did not sell Marxism to the Koreans, they sold land redistribution, better working conditions, feminine equality, and universal employment. Our own appeals were not so broad. The filled rice bowl is admittedly a better ploy to a hungry man than an ideological concept, but the point to notice is that just because an enemy or an opposing group uses certain appeals is no argument that we are entitled to do likewise. Whether or not we may is controlled by the basic premises on which a free society undertakes political communication. Unless, therefore, it is national policy to redistribute land and grant feminine equality, we may not use these and similar appeals just because the enemy does.

In other words, as Cottrell has observed, "The varied emphasis, the particular skills, knowledge and techniques of political communication are determined by the operating contexts. The more clearly these contexts are defined, the clearer will be our concept of what political communication will be under specified situational conditions."[5]

In terms of the "operating contexts" of World War II, the best we were able to do was some rough rules of thumb. The definitions were by no means clear and exact. The consensus of operators in that war was:

> Truth is the psywar operators' most effective weapon. Gear your output to as specific a group as possible. Play up human interest; play down abstractions. Take advantage of natural schisms within the target group. Play down your own position; over-extend the enemy's. Adapt format to that of target.

These guides resulted from pragmatic experience, and they are hardly sufficient for doctrine. Even now it is not certain that we are in a position to formulate guides or rules that are much more definite. As Schramm says, "The appli-

[5] Notes to the writer from Dr. Leonard S. Cottrell, Jr.

cation of [the most relevant evidence and theory from social psychology and the other social science disciplines] to the practice of psywar is not like applying a scientific formula to physical quantities. The application is partly science and partly, in very large part, art."[6]

To some extent this must be because of the almost infinite resources available to the propagandist. The means he will use to convey his messages may be numbered in digits, but it would be a rash man who would attempt to limit the possible arrangements of them. The leaflet, the loud-speaker, the transmitter, the graphic art—these, singly or combined, in one form or another, are the basic and primary means of communication. But what can be done with them, and done with them in combination with other means, seems difficult, if not impossible, to catalogue.

The danger of inflexible rules may be illustrated by considering just one medium of communication—the leaflet. There is general agreement that the propaganda leaflet should conform to the principle of the six W's: to whom; what is to be done; when; where; why; with what results. But even this cannot be laid down with rigidity. And as yet we know very little about the size a leaflet should be, the paper it should be written on, whether it should be black and white or color. As a general guide we can say that usually it must be capable of quick reading—in other words, not too long and involved. When this has been said, the novitiate has been given his charter. It is necessary to read only the leaflet, "To the Soldier Who Marches West,"[7] to recognize that if the meager guides just set down were at all binding, that leaflet could not have been written, and a brilliant conception would have been lost. The art in political communication is still in the ascendant. Rules will be

6 Wilbur Schramm, *et al.*, *The Nature of Psychological Warfare* (Baltimore: Operations Research Office, The Johns Hopkins University, 1953).

7 See Appendix, p. 259.

broken all the time by competent craftsmen who know what they are doing, and doctrine will swaddle the inexperienced and timid and will cripple the art.

Experience in this area may be summed up as follows: Our present state of knowledge is not sufficient for anything like a rigid formulation of rules. Political communication today should be wary of doctrine, but should seek operational guides and state them flexibly enough to permit operators to take instant advantage of developments in our knowledge as these become available. It is in the doing that the theories are tested, and it is in the doing that the art form assumes clarity, conviction, and meaning.

We are on different ground when it actually comes to an operation itself. In World War II, for example, the leaflet bomb in general use was independently developed in three different theaters of war. The third theater to develop it—CINCPAC—was not told what the other two had done until after it had gone through its own separate tests. The correction of this sort of situation calls, clearly, for common sense rather than any peculiar kind of skill or training.

Psychological warfare at the time of the TORCH landings in Africa provides another kind of lesson. For that operation we sent in British and American officers, British and American civilians, British and American noncommissioned officers, and British and American privates. They were not organized, they had no plan, they were given no directives, and they had no equipment.

Everyone will know the conclusions he ought to draw. Perhaps, even, it will be said that this sort of mistake was not repeated and that our activities in this field were more efficiently organized during, for instance, the Korean conflict. The evidence indicates they were. Nevertheless, the organization tested in Korea fell down. It was determined that what the Korean operation required was units of men trained for particular tasks—radio broadcasting, leaflet-writing, etc.—to be assigned where their services were needed. Instead, the organizational pattern delivered units of mixed skills, not all of which were needed. A smoothly working organization that may be expected to function efficiently in this field for operations varying along the progression peace - cold war - hot war still has not been found.

At the moment, the lessons of the past seem clearly to point to our need, not for doctrine, but for common sense and flexibility while our knowledge is being increased. We can at least say, however, that we have the tools for action as that knowledge becomes available. Essentially these tools are simple and well known, although complex in operation. They are the printing press, the broadcast, the drawing board, the camera, and, in war, the leaflet shell and the leaflet bomb.

There seems to be only one rule in connection with these tools that perhaps may be stated with rigidity. It is that the tools must be controlled by the operating agency. The evidence indicates that it is unwise to rely on other agencies to use them in behalf of political communication unless there is authority to insist, when necessary, that they be used as instructed. It has already been pointed out that FIS, in World War II, found that it could not operate effectively merely by sending to the private news agencies and broadcasting companies texts of what it wanted publicized. It had to have its own facilities or have these under its own control.

The conclusions to which a student in this field is forced may be simply stated. The tools for political communication must be owned or controlled by the operating agency. The techniques to be used in framing and drawing up the agency's messages fall largely into the category of art. There are few rules for the way something to be said or written should be prepared. The time has not yet arrived when we can speak of a doctrine for political communication. The requirement is for operational guides that are flexible and permit the creative instinct of a craftsman to flower.

There is, however, another sort of technique that the practitioner in this field wants when a program is undertaken, and it may be broken into two categories: that which helps us to get the knowledge needed to design and plan a propaganda operation; and that which helps us to assess how effective the propaganda has been to its target.

The techniques available for this are relatively few, and no one of them is sufficiently reliable to stand on its own feet. In the first place, it is difficult enough to determine the state of public opinion in one's own country, let alone its state in another.[8]

Empirical experience occupies an important place.[9] Nevertheless, we have reached a point in knowledge and experience where trial and error are not so necessary to future operations. However, we are still a long way from proceeding with certainty.

We have to keep in mind that political communication is not something to be conducted like an advertising business.

[8] The keynote of the Iowa Conference on Attitudes and Opinion Research in 1950 is still applicable: "It is not expected that easy answers will be forthcoming to the difficult problems of how mass opinion, prevailing attitudes and impending behavior may be measured. The problems, complex in the extreme, are still before those whose responsibility is to restudy them with renewed zeal. Are there better methods suitable for the peculiar requirements of public opinion measurement? How can theory be tested in the crucible of actual use? These and a multitude of other problems of broad scope in social service constitute the constructive objectives that need full and unhurried discussion by those who can contribute with factual material, knowledge and research experience." "The Aim of the Conference," in Norman C. Meier, *The Polls and Public Opinion,* the Iowa Conference on Attitudes and Opinion Research, 1950, sponsored by the State University of Iowa (New York: Henry Holt & Co., Inc., 1949).

[9] See Edward W. Barrett: "Those of us specializing in international persuasion soon found it to be an amazingly intricate business, with techniques and emphases varying from country to country. In each area of wartime operations we made mistakes, we scored successes, and we learned lessons." *Truth Is Our Weapon* (New York: Funk & Wagnalls Co., 1953), p. 34.

We ought not to be concerned with whether people like us and want to be our friends, except secondarily. Political communication is concerned with "the greatest problems on whose solution our fate depends,"[10] and this means much more than trying to persuade people to like us.[11]

Among the disciplines concerned on which we must rely for knowledge in this field, the social sciences hold an important place, even though, as yet, they are in many instances unable to speak with precision. They have, nevertheless, made great strides in recent years. We are today by no means dependent solely on sampling, attitude measurements, and the like for our knowledge of how to go about talking to people. Anthropology, political science, and economics, for example, possess powerful tools with which to give us increased understanding of different cultural patterns and the probable results of varying courses of action. The use of these tools calls for trained and experienced scientists.

The techniques, themselves, that can help us in determining whether a specific type of program in this field may be worth using can be grouped under the following headings:

- Polling
- The historical-statistical method
- Interrogation
- Institutional analysis
- Cultural analysis
- Pre-testing
- Panels
- Attitude scales
- Small-group dynamics
- Content analysis
- Situational and role analysis

10 William Esslinger, *op. cit.*, p. 72.

11 Parenthetically it may be noted that it is equally important if we want to be liked, for us to like others. We need more maturity and tolerance here. It is not essential, however, to the task of living in a shrinking world. No one will pretend that the aristocratic traditions of the South accord with those of the Midwest. But the people in both these sections of the

These techniques are not definitive. Nor may any of them be relied on independently of others. They reinforce each other. De Grazia points out that "There is no one simple method of evaluation whereby the effects of the . . . media can be accurately gauged."[12] Indeed, he warns that when it comes to assessing an operation directed towards an enemy, even such techniques as we possess cannot give us much help. Furthermore, there is no standard procedure for attempting to determine how effective we may be.[13]

It may well be asked why, if our techniques cannot give us more assistance and certainty than this seems to imply, we should rely on them at all, and perhaps even why we should undertake political communication in the first place. The first chapter of this book discussed the question of effectiveness. The evidence indicates overwhelmingly that it works. If this is the right conclusion, there can surely be no question of the importance of seeking scientific support for much of what is, so far, largely empirical knowledge.

Simply because we seem to have the right to say with considerable certainty that political communication can be effective, there is a self-evident requirement for seeking to *measure* this effectiveness, for seeking to determine with more precision than we now possess what elements are more effective than others, what combinations will yield results

country manage to do business with each other and get along under the same government. The factors that make this possible are, in essence, the factors we should look for and attempt to utilize in our political communication to peoples beyond our own borders.

12 Alfred de Grazia, *et al., Target Analysis and Media in Propaganda to Audiences Abroad* (Baltimore: Operations Research Office, The Johns Hopkins University, 1953), p. 62.

13 "In short, the question that is of most interest to the psychological warfare operator, namely, that as to the target's intellectual and emotional responses to his messages, is still largely unanswered by students of the human sciences." Furthermore, there is no standard procedure for attempting to determine this effectiveness. "The methods used to evaluate media effects vary with circumstances." Alfred de Grazia, *ibid.*, pp. 55, 56.

more quickly than others, what skills and knowledge and organization can be expected to give greater efficiency and effectiveness.

The body of scientific and technical data in this field—and it is now considerable—indicates in no uncertain terms that at the moment we are in no position to lay down elaborate ground rules for the conduct of political communication. Precise, scientific knowledge is scanty. Even so, we have demonstrably advanced since the close of World War II. And it is clear that much can be attempted on the grounds of direct observation and the application of common sense. Our military posture and strength, for example, may appropriately be used as a means of political communication in the form of maneuvers, fleet visits, and fly-overs.[14] These can well become techniques of operation as valuable, or more so, than the leaflet or the broadcast if used with imagination and skill and not as means of saber-rattling. But where common sense and imagination direct the application of this technique, the techniques of evaluation may be applied to estimate some measure of its effectiveness, and we may expect our knowledge to be increased.

It seems safe to say that common sense and imagination will always be needed in political communication, however we improve our knowledge and techniques, and it will probably never be possible to put down in a book a collection of maxims and guides and techniques that will obviate an operator having to do some thinking for himself. There will always be room for the inspired treatment of an opportunity to reinforce action by apt and deft political communica-

[14] Note the article by H. H. Martin in *The Saturday Evening Post* for March 2, 1957, titled: "Cat Brown's Kittens Have Claws; Sixth Fleet." Quite clearly, in and of itself, and by its very presence in the Mediterranean, the Sixth Fleet is in fact undertaking, among other things, a program of political communication. The question of interest is: To what extent is this realized and being capitalized on by government?

tion and usually, when the opportunity presents itself, the technique will become apparent to an alert staff.[15]

The techniques that have been mentioned so far all have to do with obtaining information about the target. There is one other aspect of political communication that needs evaluation. This is the effectiveness of an on-going operation at any given time in terms of what is being done by our own organization. The techniques for this belong in the field of administrative analysis. Only one full-scale evaluation of an on-going operation has been made in this field. This was the evaluation of our psychological warfare program in the Korean war, and it was undertaken by the Operations Research Office of The Johns Hopkins University. It provides a model for future assessment of this sort. Its method and findings have withstood critical inspection, and it clearly shows how an important tool may be used to increase our knowledge of how, organizationally, activities in the area of political communication may be usefully assessed. Unfortunately, this detailed, skilled, and disciplined evaluation, complete in itself, is the only one in its field. Applied to USIA, the tool could be expected to provide an important yardstick by which to determine how effectively organized this agency is.

These two aspects of effectiveness should not be confused.

[15] Robert E. Sherwood gives an excellent illustration of this in his book *Roosevelt and Hopkins.* "Hopkins at this time [1941] again became an important figure in the psychological warfare which, apart from the endless Battle of the Atlantic, was about the only form of combat between the British and the Germans. The BBC European services advertised Hopkins' peregrinations constantly in broadcasts to all of the occupied countries of Europe as well as to Germany and Italy. His own radio speech was translated and broadcast in many different languages and printed in leaflets distributed by the thousands and millions by the RAF Bomber Command. It took a great deal of ingenuity on the part of the British propaganda experts to explain just who Hopkins was but they managed to convey the implication that he represented the vanguard of an enormous American Expeditionary Force." Robert E. Sherwood, *Roosevelt and Hopkins: An Intimate History* (New York: Harper & Brothers, 1948), p. 322.

There is the effectiveness of one's own organization, and this can be determined at first hand, for direct access to all necessary information is possible. Then there is the effectiveness of the operation in terms of the audience to which it is directing its messages: Are these messages getting through and can their effect be measured?

Four questions are of significance here. The first duty of political communication is to draw attention to action taken. This is directly related to the first of the four dimensions of measurement that are of concern: Are we delivering the message? This is the easiest and the simplest factor to measure or determine. It may be stated: Political communication must notify its target of action taken. The measurement problem, in this connection, is concerned with how completely the notification has been carried out.

Second, political communication is required to inform its target of the significance of the action. This, in turn, is related to the second aspect of measurement: Is the desired audience attending to it? Here, again, we can measure satisfactorily. We are able to obtain good evidence, though not as good as for message delivery.

Third, political communication must give the target the meaning and interpretation of action in terms that it can understand. The third aspect of measurement here is concerned with the answer to the question: Is the message being understood and accepted? This is a harder question to answer than the first two. It is involved with what is happening inside an individual. We must rely on inferences. Nevertheless, these inferences are not negligible. If risks are taken in order to hear the message, or to pick up leaflets, we are justified in drawing a number of useful inferences. There are many acts that give us reasons for concluding whether or not the message is being understood.

Finally, political communication is required to indicate a response on the part of the target that is acceptable to both the sender and the receiver. The related question for measurement: Is the desired action being taken? This is the most difficult question, because human behavior at this level is hard to measure. It is difficult to prove causality. A simple example of this is sabotage.

These four aspects of what it is we desire to measure relate directly to the functions of political communication and concern both the operator and the researcher. They provide guides that should keep us on the track in our research so that we may be sure we are putting first things first.

This brings us squarely to the problem of research in this field. An active operation, especially when it includes a radio program, requires a solid research backing. But research accomplished, alone, is not enough. The results must be given to operators in terms that they can use.

Often this research may be in terms of logical thinking on known facts, rather than experimentation. As Davison says, "the historical record may sometimes prove deceptive . . . nevertheless it enables us to make sounder judgments in an area where progress is to be seen not as the establishment of definitive rules, but as the gradual ability to make progressively better guesses about the most appropriate action to be taken under given circumstances."[16]

The first question to be asked of research for an on-going operation in this field is: What value will this contemplated study be to the operator? Second: Does this study concentrate on the first and most important issues? This, of course, raises the question: What *are* the first things?

The research presently underway would indicate that there is little unanimity of opinion on this last question. Yet, in fact, greater unanimity exists than the on-going research

[16] W. Phillips Davison, "Some Observations on the Role of Research in Political Warfare," RAND P-226 (Santa Monica, California: RAND Corporation, 1951).

seems to indicate. One reason for this undoubtedly is that operating agencies in this field do not know what is being done outside their own boundaries.

Dr. Leonard S. Cottrell, of the Russell Sage Foundation, has had a better opportunity than many to assess the status of research in the field of political communication. He served as chairman of an advisory group directed to make a report in this field to the former Research and Development Board of the Department of Defense. It is his opinion that:

> among the military and civilian agencies interested in PW [psychological warfare], there is a real need for continuing machinery for exchanging information and for combined planning. It is very likely that this need exists also in the Joint Plans Division of the Chiefs of Staff. It has been difficult for State to cooperate with the various military services. This may be due partly to differences in point of view as to research priorities. People in the various services who had had responsibility for research have had to keep their eyes on what their own operations and planning officers wanted, and less on what other services needed. . . . Another point of difficulty in communications . . . lay in the fact that operations offices in charge of plans had no machinery to get together. Up until the time it was superseded by OCB, PSB did not achieve an integration of programs and plans for interdepartmental PW affairs.[17]

There is a striking unanimity in the tone and substance of virtually all the comment existing on the subject of research support for our propaganda operators, and the only logical conclusion is that our left hand is not properly aware of what our right hand is doing.

The criteria governing research for our activities in the

[17] Dr. Cottrell in an interview with a staff member of the Operations Research Office, The Johns Hopkins University, recorded as a Memorandum. June, 1954.

field of political communication may be given in the form of questions:

Is the problem researchable by proven method?
Is there good prospect of adapting a proven method?
Are the required data obtainable?
Is competent personnel available?
Is it possible to articulate into growing structure?
Does the aggregate program leave major factors unexamined?
What will it cost, in time and money?
Will the conclusions serve to guide action?

We are not going to have an effective, co-ordinated, significant organization for political communication without research in support of it. Much of this research must be under the direction of skilled social scientists. The funds for this work are woefully inadequate.

In 1954 less than one-half of 1 per cent of research funds spent in the federal government were devoted to social science research. Since then funds have been cut back. Yet we are firmly committed to political communication as an instrument of state. This instrument possesses powerful potentialities. Research is one of the key tools by which it must fulfill these potentialities. We are not providing the means for this research to be effective. The funds required are insignificant compared with the appropriations annually given for research in nuclear weapons.

An increase of a few tenths of a per cent, which could be relied on for a five-year period, gives every indication of yielding results proportionately far greater for the money than in almost any other line of research we are doing for our security. Considering the likely consequences of a general war, and considering the logic for the position that effective political communication is a primary instrument for its prevention, common sense seems clearly to indicate the worth of providing an adequate opportunity for research specialists in this field to determine what the disciplines of the social sciences can do if supported.

Tools and organization alone, however, will not be of much use without skilled personnel. The importance of trained men is thoroughly documented and attested. The availability of such men is another question. During the Korean war, for example, trained men were available, but not used, for a complex of reasons, none of which is particularly creditable.

In the two world wars, manpower untrained in this field had to be recruited and trained. If we were to find ourselves involved in war again—a war less than general in its aspects and therefore one in which, perhaps, the most powerful nuclear weapons would not be used—it is still questionable whether trained manpower would be available in sufficient quantity in the right places. The indications are that, in spite of all the evidence, we face the possibility, in such a contingency, of yet once more struggling to influence the minds of men with inadequate provision having been made to allot personnel.

There are, at present, three agencies charged with conducting phases of political communication: the United States Information Agency; the Military Establishment; and the Central Intelligence Agency. It is reasonable to suppose that in the event of war all three would expand their activities. This would bring an immediate demand for manpower. These agencies would be competing for this manpower with other agencies, which could use men with some of the same qualifications called for by political communication.

It is well, in any consideration of men, to keep in mind the straightforward comment of a Senate committee: "The ultimate success or failure of the information program, irrespective of its departmental lodgment depends on skilled administration and personnel. With these, the program can succeed within or without the Department of State. Without these it will fail. . . ."[18]

Political communication demonstrably calls for diverse and peculiar skills. An attack on the minds of men self-evidently is not boy's work. Throughout recorded history it has been the task of men—trained men, mature men, dedicated men, singly or grouped in private organizations within the framework of their nations—until we reached the birth of the League of Nations. It is now being considered as a national undertaking. It is not logical to expect that, as a national undertaking, it can be attempted by less trained and dedicated minds merely because it is national.

The problem is bound up with the characteristics of many of the men required—real or imaginary. Practical men of affairs tend to look askance at propagandists. The head of the OWI operation in the CBI Theater has written, "There was the recurring mistrust of the kind of people engaged in propaganda, professional users of symbols—writers, journalists, artists—probably radicals, anyway not regular."[19]

Munson says, "emotional force, creative imagery, a streak of imagination are pre-requisites for a first-rank propagandist."[20] Lockhart, who ought to know, says, "Every good propagandist must possess the qualities of a prima donna or . . . be born with one eye."[21] In writing about his own staff,

[18] Report of the Committee on Foreign Relations, Overseas Information Programs of the United States (Hickenlooper Report), Senate Report No. 406, Eighty-third Congress, First Session (1953), p. 25.

[19] Ralph Block, "Propaganda and the Free Society," *Public Opinion Quarterly,* Winter, 1948-49, p. 681.

[20] Gorham Munson, *Twelve Decisive Battles of the Mind: The Story of Propaganda during the Christian Era, with Abridged Versions of Texts that Have Shaped History* (New York: Greystone Press, 1942), p. 90.

[21] Sir Robert H. Bruce Lockhart, *Comes the Reckoning* (London: Putnam and Co., 1949), p. 155.

the Political Warfare Executive, he talks of "their enthusiasm, essential in propaganda work."[22] Linebarger emphasizes the importance of ingenuity in work of this kind and decries its conventionalization.

Men with the characteristics thus expressed or implied do not lend themselves easily to stereotyped procedure or gracefully become part of an operation that depends on work being done "by the numbers." The answer is not to dispense with such men. It is obvious that certain phases of political communication require the skills and talents of people with these characteristics. The question is one of how most effectively to organize an operation in which such men are important components.

The rules for making an organization of prima donnas tick are not yet susceptible of scientific measurement. The handling of men with these qualifications is more an art than a science. The trick is to get talent to produce in an age that lives by technology, without recourse to the devices technology sets up in order to keep the machine running. This is an art, and the skilled administrator of talent comes by his skill only after pragmatic experience, plus an initial interest in people and an understanding of their foibles. Writers, for example, are unlikely to produce if they must punch a time clock. They can appreciate a deadline.

Furthermore, in this field, talent is restive under mere "orders," necessary though orders are. It is impossible to "order" a writer, in a free society, to write a leaflet. The better writer he is, the less he is likely to write well without being in general agreement with the objectives the leaflet is designed to achieve. Doob, at the conclusion of his experience with OWI in World War II, observed that "you could lead an operator to a directive, but you could not

[22] *Ibid.*, p. 215.

make him implement it—no matter how strict the system of controls—unless he understood it, approved of it, and/or had trust in the policy-maker."[23]

The Clark-Darlan agreement in World War II affords interesting confirmation of Doob's finding. Although this agreement affected our psychological warfare operations in the North African Theater, it appears that our personnel in Algiers were not informed of it. When the agreement became known, their first reaction was that our principles had been sold out. For a long time world opinion was shocked. And our propagandists were shocked—and therefore ineffective—along with it. It is worth keeping in mind: Propagandists, if they are any good, may not be "ordered" to propagandize in a free society.

The reasons for this are very clear and most logical once the premises on which these propagandists are at work are understood. Intellectuals—and it is intellectuals who are the creative backbone of political communication—possess a responsibility to themselves and to their era to express, portray, vitalize, and criticize the currents stirring the thought patterns of the age in which they live. They may not abdicate this responsibility without a certain loss of personal integrity. This is a further underlying factor supporting the necessity for a free society's conducting political communication on the principle of truth. If it appears that this principle is being violated, the propagandist in a free society is among the first to whom the reasons for a given course of action should be explained. Otherwise, as the history of the Clark-Darlan agreement so clearly illustrates, he "dries up." Counterpropaganda, in such circumstances, is left to run down the field with little effective opposition.

[23] Leonard W. Doob, "The Utilization of Social Scientists in the Overseas Branch of the Office of War Information," in Daniel Lerner (ed.), *Propaganda in War and Crisis* (New York: George W. Stewart, Inc., 1951), Ch. 17, p. 312.

It is one thing to agree on the need for qualified personnel; it is another thing to agree on what the qualifications are. There is no one source of supply. Good propagandists come from many sources. As yet there does not seem to have been any thorough analysis of the ingredients that, if they are present in an individual, allow us to determine with some assurance that he will make a good propagandist.[24] In the main we must still lean to a considerable extent on the judgment of men with experience in this field when it comes to selecting personnel. Men who are intelligent, and themselves trained, apply in assessing their fellows what we may call rules of thumb, based on their own experience, on the degree to which they have developed the art of evaluating human nature, and on their skill in estimating the character of men. Certain backgrounds, we know, are of value, but they are not indispensable. Certain professions are more likely to yield the desired characteristics than others, but, again, this is no sure guide.

In World War II the British drew widely on their society for their propagandists. They came from all walks of life. Lawyers, farmers, professors, newspapermen, stockbrokers,

24 RAND has noted: "The study of the backgrounds, personalities, and skills of political propagandists is an important part of the field of propaganda communication. It is hoped that ultimately an analytic literature will be developed on the subject of 'symbol specialists.' In the meantime, relevant data must be gleaned from (1) more or less pertinent theoretical and experimental contributions of psychoanalysts, social psychologists, and sociologists specializing in fields like the dynamics of leadership; and (2) discriminating biographies of individual propagandists, statesmen, and politicians."

Chitra M. Smith, *et al.*, Bureau of Social Science Research, The American University, "International Communication and Political Warfare, An Annotated Bibliography," RAND RM-974, October 23, 1952. Out of print. Superseded by Smith, Bruce Lannes and Chitra M. Smith, "International Communication and Political Opinion," a guide to the literature prepared for The RAND Corporation by the Bureau of Social Science Research, Inc., Washington, D. C. (Princeton, N. J.: Princeton University Press, 1956).

psychologists—all were represented in Great Britain's PWE. And of this motley assortment Lockhart says, "I do not think that any one profession provided any initial advantage of training. A propagandist is born and not made."[25]

A similar motley characterized our own efforts.[26] There is no one source of supply for propagandists; their qualifications are legion; and as yet, we have no scientific method of determining in advance who to summon to this task. Nevertheless—"Without skilled personnel we shall fail."

If we do not know from where we are going to draw our specialists for the work of political communication, it must follow that there is a definite need for training men who come from diversified backgrounds. A national operation in this field is defined by certain boundaries, premises, and assumptions. These must be understood if there is to be any hope of success. Furthermore, in addition to inherent qualifications and skills, there must be training in the application of these talents.

Essentially, two kinds of training are necessary. There must be knowledge of the component parts that make up the spectrum of what is called political communication and of how they fit together. For example, the kinds of intelligence necessary and how intelligence supports policy, the kinds of liaison necessary among departments of government, and the relationship of policy-making to the explanation and interpretation of that policy.

This sort of knowledge can be explained and its value demonstrated in training courses. It is hard-won knowledge, some of it gained in World War I and then lost. Fresh knowledge was gained in World War II and old knowledge recaptured. It, too, was lost or hidden before the Korean war. We have now still later knowledge, and

[25] See Sir Robert H. Bruce Lockhart, *op. cit.*, pp. 156-57.

[26] See Daniel Lerner, "Introduction," in Daniel Lerner (ed.), *Propaganda in War and Crisis*, p. xii.

much of the old knowledge has been set down in transmissible form, whereas in World War II and immediately after, it had to be transferred by example and word of mouth. This sort of knowledge can be given potential specialists in training courses.

The second sort of knowledge for which training is required is never completely learned. It is compounded of experience, disciplined acquisition, and the possession of a certain gift for working in the area of ideological conflict and growth. Colloquially, in essence, it means understanding what political communication is all about.

A certain amount of this kind of knowledge can be given in training courses. A considerable portion comes only with practice. Some of it comes from studying the history and experience of past operations and testing the conclusions against present situations and the increasing sum of our knowledge.

It must also be taken into account that men engaged in this profession will need to make considerable adjustments from the prevailing attitudes common in the business and professions marked by private enterprise. News, for example, may not be written and explained from the viewpoint of the individual writer. It must be reported through the eyes of the ministers of government, for it is their intent behind the policy they make, and the actions they take, that is significant. And it is this that must be interpreted, not the subjective viewpoint of the individual.[27]

This sort of adjustment can be made considerably easier, and the resultant output of work considerably more effective, by training. At present there is very little training for

[27] See Charles A. H. Thomson where he expands on the importance of "reporting and explaining the news not as the reporter saw it, but as the President, the armed services, or the Department of State saw it." *Overseas Information Service of the United States Government* (Washington, D. C.: The Brookings Institution, December, 1948), p. 32.

the men who enter this field to become specialists, except what they get on their job.[28]

The only training course given in the free world today is given by the United States Army at Fort Bragg. It lasts six weeks. It is not adequate, and the Army knows this. But it is an honest attempt. And there is no other such course. USIA, with a daily operation on its hands that is continuous, does not have one. The Army, which possesses a specialized and far narrower interest, has seen fit to make the attempt. There is considerable food for thought in this.

Several years ago an attempt was made to institute a far wider course under the aegis of the Department of State to which men from the departments of government would go. For six months program after program was drawn up. A realistic curriculum was envisaged. It encompassed such essential subjects as: theories of international politics; the practice of United States policy-making; a study of the methods by which the United States determines its foreign policy; government organization for the conduct of foreign policy and foreign information.

Nothing now remains except the skeleton of this attempt, but the dry bones, if the right wind breathed on them, could be brought to life again and form the nucleus of a forceful, adequate, and important training program for specialists in political communication and other officers of government with interests related to the field. At the time it was projected, interdepartmental differences and bureaucratic frustrations combined to prevent its realization. On the cerements that, in a government file, neatly cover the body, it is laconically written that the Navy, the Air Force,

[28] Edward W. Barrett, former Assistant Secretary of State with responsibility for our information program, has observed: "There is now no real training center for young workers in the complex work of international persuasion. Government agencies generally must recruit young men having only a few of the skills and attributes needed; they learn by osmosis while they work." *Op cit.*, p. 285.

and the office of the Joint Chiefs of Staff opposed the plan; that the Army, like the guests of the King who gave a marriage feast for his son, "made light of it"; that the Department of State, like the Church at Laodicea, was neither cold nor hot.

There are other agencies that wrote pious valedictories for the interment ceremony. Each man cut the baby's throat —so well and skillfully that it is doubtful if the little corpse ever knew it was bleeding. And today there is no training course at all adequate for our needs. Our political communication is inept and weak when it should be decisive and powerful. It has been inept because it has not been a planned accompaniment of our actions, which have met with an undeniable measure of success. It has been weak because we have failed to act on the premises governing its use. The initiative has rested with the enemy.

We have the tools. We know the requirements. We can supply the men. There is no training for them as yet, and no real attempt to put this instrument where it belongs so that it can speak with a powerful, national voice in defense of free men and their aspirations.

The weapon is on the wall. It belongs in our hands, for the house that we have builded is under attack.

VII The Requirements if the United States Is To Develop the Instrument Effectively

The first and most important condition to be met before political communication can be effective is that government must act *as if* it believed the instrument possessed the capabilities that a study of its potentialities indicates are present. This belief has been noticeably lacking in the past. It is lacking today. As long as government intimates by its manner that it shares the doubts of many about the efficacy of the activity, it matters little what government may *say* about it. Lip service, which is one of the chief things political communication has been accorded so far, effectively holds back the full potential of the instrument. There must be much more than lip service. It must be clear not only that government thinks political communication is important, but that in its day-to-day operations it treats it in such a way as to show that it is important.

How can government do this? In much the same way it shows that the Department of State is significant and vital and that the Department of Defense is imperative and a key department. The same kind of attention accorded to political communication will quickly demonstrate its position in the government hierarchy.

There must be leadership if this position is to be attained. In all probability the kind of leadership required can come only from the White House. It is by no means a matter of considering only the facts and the logic that testify to the importance of political communication. By definition it is pre-eminently concerned with political realities and therefore subject to political considerations. Politics is not yet a science. Therefore facts and logic, while necessary, are not enough. The exigencies of political life must be taken into account. The only figure in the United States who can do this in terms of national objectives, overriding localized political prejudices, in all probability is the President.

At this juncture in our history, political communication is a prime requisite if for no other reason than to inform people of the nature of the technological developments and the strides in knowledge that have marked the past decade. If, now, we fail to make clear to the minds of men, not only in our own country but throughout the world, the nature of the world revolution of which we are a part, we must expect one of two things to happen. Either we shall, sooner or later, be forced to fight for our survival, or we shall be forced to capitulate to an ideology that, as free men, we find abhorrent.

So far, it is clear, as a nation we have been content to think of political communication in terms of propaganda and psychological warfare. And so far, in the public mind, no great attention has been paid to the activities carried on under these names. Perhaps it even can be argued that this shows the essential ability of our people intuitively to discern what merits support and what is secondary or waste effort, for it requires no very great skill to demonstrate conclusively that propaganda, by itself, is weak and ineffectual. Psychological warfare, as much of it has been practiced, is technique and trickery. It carries this connotation in the public mind. And the studied refusal of the public to be impressed by these activities shows only its innate good sense.

The larger and deeper meanings implicit in the concept of political communication have been obscured by activities in the field that, in reality, *are* secondary. At the same time it is the practice of these activities, and our successes and failures with them, together with our growing recognition of the inseparability of propaganda in peace and psychological warfare in war, that enable us to believe, with logic and some assurance of success, that the broader concept of political communication, with the activity properly organized and directed, can be of significant value to us in the revolutionary and troubled years we undoubtedly face. For this to be possible, however, the instrument must be given status and prestige and encouraged to develop its potential.

This requires action by the White House that unmistakeably establishes political communication as a coequal partner with the other three instruments of a sovereign state: the political, the economic, the military.

The Co-ordination of Policy and Operations

A second requirement stems from the nature of the activity itself. Political communication, to be effective, has to interpret and explain the national aims of the United States and clarify our truest and fundamental objectives. The point to notice is that it should reflect *national* policy. National policy is not made by any single department of government. Furthermore, because of the increasing complexity of the problems involved in decision-making, there is increasing interdependence among government departments when policy is being determined. Foreign policy cannot be made solely by the Secretary of State and neither can military affairs be settled by the Secretary of Defense without reference to other departments of government.

National policy, when it is being determined, increasingly requires the participation of the major departments of government. It is not reasonable to expect political communication to be able to do its job if its direction is placed in the hands of either the Department of State or the Department of Defense—the two departments concerned primarily with our political and our military policies. There is no department of economics—the third element of our national posture with which political communication is concerned. Our economic posture is the result of the activities of several agencies, for example the Treasury and the Department of Commerce. No one of these departments or agencies may direct our political communication, because no one of them is above the battle of departmental politics. And the history of our propaganda and psychological warfare activities so far clearly reflects the difficulty, if not the impossibility, of an agency under departmental direction being able to function objectively in relation to our national aims and objectives. These aims and objectives result from the sometimes conflicting requirements of such departments. It is the final position that political communication must explain to the world, and it is not logical to expect an agency under the control of one of these departments to be able to rise above the politics of its department after the give and take of determining policy has been resolved. It is, furthermore, necessary to interpret and explain this internal conflict that precedes policy determination, and, again, this is difficult when one's own foothold in government is in one camp or the other.

In certain respects, nevertheless, the most logical point from which to direct political communication comes under the wing of the Department of State, but the Department of State, itself, has done a great deal to make this, in practice, impossible. In the early and middle years of our concern with this activity, the Department of State looked down its nose at it. It took the position that it wanted to determine the *policy* our psychological warfare should follow, but it did not want to get its hands dirty in the actual operations. More lately it has, in addition to this attitude, compromised its ability to rise above departmental positions by the arguments it used in objecting to the creation of the Psychological Strategy Board, and by the way it went about the still unresolved difference of opinion with the Department of Defense on the question of who had jurisdiction over psychological warfare in the event of hostilities.

There is, however, a more cogent reason for excluding the Department of State from consideration as the department responsible for our political communication. The concept behind a department of state or a foreign office in a free society is based on the conducting of political negotiations among heads of states through representatives, that is, ambassadors or ministers, acting for other heads of states. Our political relationships with other sovereign entities are matters of law and the boundaries within which they are conducted are legal boundaries. Our political communication has to be addressed to peoples and to masses. And, even if the public opinion of these peoples may be said to be "sovereign" in an ultimate sense, they are not, nevertheless, legally, heads of states. In other words, the question is: Can the apparatus of law be stretched comfortably to include the propagation and discussion of ideas and mass movements? These two worlds, the world of law and the world of ideas, are not capable of being mixed effectively, if they can be mixed at all.

Secretaries of state, ambassadors, ministers—all the appanages of their work—are concerned primarily with formal communication among governments. This is their business, and it is an important and full-time task, with its own requirements for upholding the aims and objectives of a nation. These men probably cannot and must not attempt to communicate with peoples and mass movements. In all probability diplomatic apparatus is required to view with extreme caution, possibly even with distaste, involvement with the full range of actions open to political communication.

If this is so, it is a sufficient reason both for the Department of State's record of reluctance to become embroiled in propaganda operations and for keeping these operations out of its hands. Dr. George S. Pettee[1] inclines to the view that the logic of this pushes in the direction of a need for a Department of Foreign Affairs with greater responsibilities than those presently carried by the Department of State. Such a department would be a three-pronged instrument with one prong concerned with our diplomatic requirements, another with our foreign economic commitments, and the third concerned with political communication. Recent developments certainly tend to support the direction of this thinking with the establishment of the International Cooperation Administration in the Department of State. Nevertheless, if it takes place, it is unlikely to be in the near future, and it is the immediate future with which this discussion of political communication primarily is concerned.

The requirement is quite clear: administrative direction that will permit us to talk to peoples in such a way that those listening will be convinced that it is indeed the United States that is talking and not just a well-intentioned group of its citizens.

There are three agencies in our government at the present time that have responsibilities for this under certain conditions. They are the United States Information Agency, the military establishment, and the Central Intelligence Agency.[2]

1 Notes to the writer.

2 In this connection see the series of six articles, "How We Are Fighting the Battle for Men's Minds," by Anthony Leviero, in the *New York Times*, December 9-14, 1951.

These, it is suggested, must be accepted as "given." In view of the difficulties we have had in trying to establish any apparatus to carry our messages to the world, it does not seem reasonable to argue for yet another reorganization with its attendant temporary loss of direction while still another course of procedure is attempted. We shall do better if we take what we have and establish it.

This means that some way must be found to ensure that the three agencies we now have—USIA, the military establishment, and CIA—speak in directed, unified terms for the nation as a whole. At present there is little or no liaison among them either in planning for what they may be called on one day to undertake or in operations. At the time of the Korean war, for example, USIA and the Psychological Warfare Section of the Far East Command were not effectively linked, nor did they attempt to reinforce each other or to exchange views as to how best their tasks might be complementary.

The only place in government where both policy and operations can be co-ordinated is, again, the White House. There must be this central direction. Political communication must speak with unity, coherence, and emphasis. With three agencies in the field, each functioning at least semiautonomously, this is impossible without a captain for the ship. Under our form of government as it is presently constituted, this co-ordination can come only from the White House.

The Role of "Black" Operations

Black operations in the field of political communication are, or should be, highly specialized. They are made up of those activities in the field of propaganda and psychological warfare that a government orders to be undertaken behind its back. If they are detected, government, generally speaking, will, if it can, disown them. Operations of this nature are not bound by the usual rules controlling actions and messages between people and nations. They are, to begin with, essentially deceptive and intended to deceive; otherwise they would be openly acknowledged.

A classic example of black work is Operation Annie, carried out in World War II. Operation Annie's headquarters was a radio station, purporting to be German, which broadcast to Germans. During its life it was planning for just one purpose: to deceive and disrupt the German people and German commanders when Germany was invaded. For this purpose it was scrupulously careful in its preparation of the minds of its listeners. Its news was accurate. Its facts were true, and their truth could be checked. Everything it did was predicated on building up in the minds of its listeners the implication that its integrity could not be questioned and that it could be trusted completely. Then came the invasion and the opportunity for which Operation Annie had been conceived.

At the height of the invasion of the Rhineland, it went on the air with lies, distortions, and false "facts." Because of its reputation, it was believed. Because it was believed, it demoralized. A dirty piece of work, justified by the realities of war, was well done.[3]

The circumstances surrounding the broadcasting of the radio message that brought about the surrender of the Italian fleet in World War II contained elements of black propaganda. Italy was ready to surrender, but Mussolini was still at the head of the government, and the fleet was operating in the Mediterranean. It was important to make

[3] See "Operation Annie," by Brewster Morgan, *The Saturday Evening Post,* March 9, 1946. Also Wilbur Schramm, *et al., The Nature of Psychological Warfare* (Baltimore: Operations Research Office, The Johns Hopkins University, 1953).

contact with the fleet and arrange for its surrender rather than for it to fall into German hands. The waveband used for international distress signals was chosen. The Italian fleet picked up the message and surrendered at Malta.[4]

Goebbels, in his diaries, describes an attempt he made in May of 1942 to deceive the Russians about the next major German offensive, although evidence of success is lacking.

> For certain reasons we launched an "unauthorized" article in the *Frankfurter Zeitung* which discusses the economic and operational possibilities of an attack on Moscow. With this article we are trying to divert the attention of the enemy to a different sector from the one in which we actually intend to attack. Whether we shall succeed in getting the Bolsheviks to fall for this is very doubtful.[5]

Black propaganda methods also played their part in the armistice negotiations with Italy. In order to convince both the Italians and the Germans that Italy was taking herself out of the war at the time of the Salerno landings, a speech by Field Marshal Pietro Badoglio over the Rome radio was considered essential. The text of the speech Badoglio was to make was in the hands of Allied headquarters at Algiers. At the last minute, Rome notified Algiers that Badoglio could not broadcast. With the text of Badoglio's message in Allied possession, Algiers went on the air and compromised the Field Marshal sufficiently cleverly and sufficiently sympathetically for him to be persuaded that it would be better for him to read his own text. The ruse was successful, and Badoglio went on the air. Italy changed sides, and the Salerno landings were unopposed by Italian troops.[6]

[4] William E. Daugherty, *A Psychological Warfare Casebook* (Baltimore: The Johns Hopkins Press, 1958), "Surrender of the Italian Fleet—1943" by Edward M. Kirby and Jack W. Harris, pp. 408-410.

[5] *Ibid.*, "Propaganda for Strategic Deception," by M. J. (Morris Janowitz), pp. 381-82.

[6] *Ibid.*, "An Armistice Message to Badoglio," by Richard H. S. Crossman, pp. 410-11.

Again during World War II, the Belgian underground published an imitation of the *Soir*.[7] The possible varieties are virtually endless. "The means of black propaganda are many. Some of them represent bold violations of international law and cause endless embarrassment to the enemy. Among these more violent forms are the counterfeiting of enemy currencies and the forging of ration cards and identification papers."[8]

This sort of work, underhanded, sometimes nasty, always done in the dark, calls for qualities that civilized, honorable men may not recommend to undergraduates in commencement addresses. As Linebarger says, "Fabricated sources require persons adept at illicit imaginativeness, impromptu forgery, and general devilment, . . ."[9]

What is more, the fact that activities in this field are going on today[10] reflects not only the alarming threat to our survival implicit in the revolutionary challenge the world faces, but also the tremendous technological advances in methods of mass communication, which make it possible to poison, not one wellspring at a time, but, possibly, the source of the water supply for many such springs.

We are required to ask ourselves if these underhanded machinations are justified—if the people of a free society, in reality, support them, or would support them, if they knew all that was going on. The answer cannot be a flat yes or no. In time of war trickery and black work are condoned.

[7] *Ibid.*, "Black Propaganda: Its Techniques and Requirements," by Ladislas Farago, p. 672.

[8] *Ibid.*, p. 672.

[9] Paul M. A. Linebarger, *Psychological Warfare* (first ed.; Washington, D. C.: Infantry Journal Press, 1948), p. 44.

[10] William E. Daugherty, *op cit.* "Black Propaganda: Its Techniques and Requirements," by Ladislas Farago. "Today there are several agencies conducting black propaganda on both sides of the iron curtain, . . ." P. 671. See also Anthony Leviero's series in the *New York Times*, previously referred to.

They are seductive operations, and they appeal to men. This is one of their dangers. There is something about them —the risks, the pitting of wits against wits with life itself as the price for failure—that makes men who have engaged in this kind of work reflect on it with nostalgic affection after their escapades are over. As a nation we had no particular penchant or apparatus for it until World War II. We learned well and quickly then. Dr. William J. Morgan, in his book, *The O.S.S. and I,*[11] has given a picture of many of the elements that go into successful practice of this sort of operation. In the situation we have found ourselves facing in the postwar world we have decided it is not possible, if we value our way of life and our survival, to discard the techniques and stratagems developed and refined in war.

We ought to ask ourselves, however, if, in fact, we need these activities at any time except during war itself. And by pragmatic tests alone, the answer has to be yes. The problem is not one of simple proportions. It is intricate, complex, and compounded of that dichotomy in our national character so well dissected by Karen Horney in her book *The Neurotic Personality of Our Time.*[12] As a nation we are exhorted on Sundays to view our fellow man with brotherly love. But Mondays through Saturdays we work in a world we know is aggressive and highly competitive. In order to win, we may not be too squeamish about our methods.

This increases the difficulty of conducting black operations by responsible means. And the requirement for responsibility in this field is a high one. What we face, in fact, is a rather dishonorable practice, permitted by custom in war, which has become necessary in a cold war and which is ruthlessly and most irresponsibly practiced by the Soviets, now becoming a recognized part of our international pro-

11 New York: W. W. Norton & Co., Inc., 1957.
12 New York: W. W. Norton & Co., Inc., 1937.

ceedings, albeit secretly and kept hidden under the rug when we have visitors. In the context and the climate of political practices in our society it is inevitable that, under such circumstances, anyone who enters this subterranean labyrinth seeks prestige and influence. And since brotherly love and competitiveness are in constant conflict in our society, this means that our good intentions, our agreements openly arrived at, our attempts to persuade and influence the minds of men honorably and fairly are in conflict with means we do not openly countenance, also designed to persuade and influence.

A major question revolves around the problem of which activity, in the field of political communication, in reality governs our attempts to bring about relevant political action from the people with whom we are in contact.

It is unlikely that we shall find a satisfactory answer to this question until we have determined the premises that govern black operations. The premises that govern our white —or open—communication to foreign peoples have been set forth in an earlier chapter in this book. Do they also act as determinants of the lengths to which we may go in our black work? This writer defends the position that they do. In other words, they have priority and are controlling. This does not mean that in addition to governing our open communications they operate to determine what we shall say or do in our secret communications. But it does mean that the premises on which black operations are undertaken must accept as their first and basic rule of operation the position that any decision to undertake a particular piece of work, in secret, must be subjected to the test of whether, if it becomes known, the damage it will do to our acknowledged aims, openly stated, will do more harm than is acceptable if we are still to maintain that we are an honorable and ethical people.

Black operations are necessary because it is an imperfect

world. But the knowledge ought not to mean that we accept the requirement as of greater validity than the precepts we have established for our open relationships with each other. The implications inherent in undertaking black operations at all have not yet been thought out and fastened down. The role of the established agencies of government in clandestine operations has not been clearly enunciated. Until it is we are in danger, the danger of having our operations in this field backfire in our faces. Should this happen, it will be the President of the United States who will be held responsible. There is no need for it to happen, but if it is not to happen, there is a pressing requirement for the relationship of this activity to the open and honorable aspects of our national life to be unmistakeably pinned down. And this must be done by responsible and practiced students of government. It is not reasonable to suppose that the characteristics of "illicit imaginativeness and general devilment" will accomplish it.

The premises that ought to govern the black propaganda operations of a free society have not yet been codified. There is a requirement for this to be done. Moreover, it should be done openly and the premises subjected to public scrutiny and debate. There is nothing secret about them. The work they govern is secret, often highly so. But the premises may not be secret. In a free society it is the right of a people to know on what terms its government may undertake, in the national interest, to trick, distort, suppress, arouse fear, or, at times, simply convey straight facts, but in a manner that keeps their significance from being realized by the general public. In fact, in a free society, in the last analysis, it will be public opinion that determines the validity of the premises.

Whoever, either as an individual or as a group, undertakes the coding of these premises in all probability requires operating experience of the activities that fall under the heading of black operations. Theory alone, in a field that is not susceptible of precise measurement, will not suffice. It is important to know and understand also the practice of the activity.

Until this coding is done, it will be difficult for those with concern in this field to speak with authority. Nevertheless, there are reasonable grounds for assuming that certain points in dispute and certain kinds of relationships will be clarified. One of these centers on the nature of the kinship between black and white operations. Is there, in fact, a relationship between them? In the field of operations the answer is probably no. There are, however, certain pragmatic facts that link them. For instance, the intelligence that is of value to one is quite likely to be of value to the other. In a surprisingly large number of instances it indubitably is. And this is natural enough and to be expected, since both black and white are concerned with aspects of political communication. If this is so, then it is pertinent to determine who is to decide what intelligence the operators in both fields are to have. In practice, there is a natural affinity between operators in the field of black operations and the intelligence fraternity. Both are secretive by nature. White operations, on the other hand, want intelligence on which to base methods of procedure and decisions that will be openly apparent. This does not mean that white operators may not be trusted with secrets. But, in practice, again, and from the historical record, there has been a tendency to hold back from the operators in the open, responsible field of political communication the facts and intelligence they must have if they are to be effective.

It is also pertinent to ask what it is we expect from black operations. How effectively can we communicate with alien societies in secret and by deception? On what grounds can we expect this kind of work really to win friends and bring about relevant poltical action when the government officially will not openly acknowledge responsibility for the actions undertaken? Merely to argue that it is practiced by Russian leaders in their relationships with foreign peoples does not give a free society license to ape the model. The history of ideas clearly shows that lasting effects are gained when the force and logic of the ideas themselves are responsibly and openly debated, analyzed, and presented. In the long run secrecy defeats itself. It is the facts that count, and they count only when they are openly available for scrutiny and testing.

Covert operations are not always deceptive. There are times when government may wish to communicate information that is legitimate and important, but that it is unwise to publicize openly or to convey through the conventional secret channels of diplomacy. For example, if intelligence leads to the supposition that an enemy is making certain plans that are based on an erroneous estimate of our ability, our will, and our means to act, and that it is to our interest to frustrate, a clandestine channel may well be the best and proper one to use. For one thing, to state openly that we know of enemy plans may give away an important intelligence source by revealing that a leak has occurred. Second, no enemy will, in reply, openly admit to such a plan, and we are left in the position of making a to-do about a situation that is denied. The covert approach may well be the best in circumstances like this. The necessary deterrent information will be conveyed; both sides will know the proper meaning to be given to communication. But neither can deny to the other a course of action about which both will profess complete official ignorance.

It would seem that there may be a corollary to this. It is difficult to envisage a continuing need for black operations except against an enemy or potential enemy. We should be able to talk openly and with candor to friends and neutrals.

The key to black work lies in knowing which clandestine

operation is legitimate and which is not. The answer is not going to be found in a book of rules. Each instance has to be decided in terms of relative values, aims, objectives, and situations. This brings the question down to a matter of a nation's character, innate decency, and integrity. This makes the requirements for responsibility in this field the more compelling. When no precise rules can be formulated, and when the activity itself is questionable, when it involves the possibility of compromising a nation's honor and its announced objectives and deepest ideals and aspirations, it is all the more important for its operations to be in the hands of mature, responsible, and level-headed men. Covert operations are a heady wine, by their very nature. Their work inevitably involves an apparent disavowal of the ordinary rules of conduct governing relations among men. In fact, this disavowal is apparent only. It is not real. But it may easily become real if irresponsibly discharged.

We, in this country, have no long experience and acquaintance with this kind of work. There is all the more reason, therefore, for examining the premises on which such work may be undertaken and formulating these premises in such a fashion that they serve as guardians of the nation's real and essential values and beliefs.

A Permanent Congressional Committee

It is obvious why a free society will always have difficulty in conducting political communication. The voice of a free people is not one voice; it is many voices. A dictatorship labors under no such handicap. But there is a certain air of sameness about authoritarian propaganda that is reminiscent of a needle that sticks in a groove in the gramophone record. However, there is no need for the messages a free people may wish to send to be more confused and complicated than they have to be. And it is beyond dispute that so far we have not done very much to ease the problem.

There are good reasons for considering that a permanent Congressional committee on political communication could materially assist in lessening the confusion that surrounds our attempts to persuade others to support our aims and objectives. Reduced to their essence, these reasons come down to the right of a free people to know what is being done in their name. If all our activities in the field of political communication were open and legitimate, there would be no need for such a committee. These activities could be subjected to legislative scrutiny at any time public opinion was sufficiently aroused to demand this. The fact is, however, that a number of the things we do in this field are secret and properly so. Few people know what, in this secret field, we are attempting to do to the minds of men in the name of the American people. We do not know the principles on which black operations are undertaken nor the premises on which decisions in this area are made. There is, at present, no sure control over such operations, and a free society possesses an inalienable right to know what its representatives are saying and doing in matters that affect its character, prestige, and honor. One of the "don'ts" in Radio Free Europe, for example, has been given as: "Don't lie unless you are completely sure that there is no possible way for the listener to discover the falsehood."[13]

The right of a free people to know what is being done in its name is in open conflict with the necessity for secret operations. These operations cannot be submitted to public scrutiny. Many of them are essential, given the pragmatic

[13] "A Thematic Analysis of Radio Free Europe and Voice of America Broadcasts to Albania and Bulgaria and an Examination of the Manner in which the Production of Radio Propaganda is Controlled." Robert Henri Binder. A thesis presented to the School of Public and International Affairs, Princeton University, April, 1953. P. 63.

conditions of our world. So the question is: Can provision be made for ensuring that secret operations are conducted in the national interest? Much the same sort of question came to the fore after we had developed the atomic bomb. A solution similar to the one we applied then appears logical here.

> We have developed a method for exposing the most secret operations of the government to the representatives of the people on a basis which might be called trusteeship for public information. The Joint Committee on Atomic Energy has been able, generally speaking, to hold the confidence of the military and the scientists. It has also been able so to establish and hold the confidence of Congress, and of the public, that even the Atomic Energy Commission is under the minimum essential legislative scrutiny.[14]

Some sort of similar joint committee could serve the nation well at this moment in our history. Its membership would require careful thought, for it would have to pass the same test faced by the atomic energy committee: It must possess the confidence of both government and people. If it possessed this confidence, it would undoubtedly act as a powerful influence on our public opinion to stabilize the roles of the various agencies with operating powers in the field of political communication. It is difficult for us to concentrate on persuasion when our efforts are repeatedly being investigated and repeatedly in jeopardy from being unable to count on funds. The investigations and the cuts and threats of cuts are symptomatic; symptomatic of a very proper uneasiness about the way we are going about our business in the field of propaganda. There is recurrent distrust of propaganda as an instrument. And, as long as propaganda is thought of in the unsavory context of modern

[14] George S. Pettee, The Walgreen Lectures, "The Soldiers and the Republic," Lecture III: "The Privacy of the State." University of Chicago, January, 1954. Unpublished manuscript.

times, this distrust rests on good ground. If we are attempting merely to ape the enemy, if we are relying on trickery and smart politics to attempt to achieve what lastingly can be achieved only by conviction, decency, and innate integrity supported by knowledge, there is little that we can really expect from propaganda. This is not the concept underlying political communication.

A joint committee in this field has been repeatedly advocated.[15] In all probability only a joint congressional committee along the lines of the atomic energy committee, selected to assure the safety of the information involved, can provide "the minimum essential legislative scrutiny."

Defense and State Liaison

An important factor that operates to control any effective political communication program we may decide to under-

[15] Edward W. Barrett, for example, says: "If a great American corporation faced unprecedented public relations problems, it would hardly expect the stockholders to have meetings, debate every phrase in past company announcements, and decide what to do. It would not even expect the full board of directors to debate perennially the complex problems of the company's public relations. Rather, it would name a small select committee of directors to work closely with company executives on the problem. That pattern may well provide the answer for the Congress. It is not an easy answer within the present Congressional framework, but it is the only sound answer that I and other veterans in this field have been able to find. It amounts to establishing a special 'Joint Committee on International Information' to act on behalf of the entire Congress in the whole field of international information, educational exchange, and psychological strategy." *Truth Is Our Weapon* (New York: Funk & Wagnalls Co., 1953), pp. 256-57. See also p. 293.

The Ninth Semiannual Report of the United States Advisory Commission on Information says: "It is recommended: That a permanent joint congressional committee composed of representatives from existing committees concerned with foreign policies and appropriations be appointed for general liaison with the work of USIA." Eighty-third Congress, Second Session, House Document, No. 311, February 2, 1954, p. 2.

See also the hearings before a Subcommittee of the Committee on Foreign Relations, Overseas Information Programs of the United States, United States Senate, Eighty-third Congress, First Session, pp. 485, 489.

take is the continuing nature of the task. If it has been begun in peace, it will continue in war. It will be modified and changed, but it cannot be stopped. Either it will continue as an organized, deliberate, and directed attempt by government, or else the actions of government, uninterpreted and unexplained, will convey meanings that may be distorted or misunderstood by those who learn of them. In either case some form of political communication goes on. If the activity is to be an effective instrument of policy, obviously it must be directed. This requires among other things that the relationship, in a changing world, between what it was necessary to do or say yesterday and what it may be necessary to do or say today be explicitly explained.

There cannot be this kind of directed, integral, and compelling continuity unless there is liaison among the key departments of government responsible for phases of our national policy. We must assume that, if war comes, the long-run aims and objectives of our policy are to be served by it. They may be modified by the overriding necessity, first, to secure our survival, but we may not assume that the war destroys them altogether and substitutes another policy. There is, therefore, a clear need for our political communication to be intimately linked to our political moves and our military posture, both during peace and in time of war.

This demands sympathetic and whole-hearted liaison from both the Department of State and the Department of Defense with whatever agency is responsible for interpreting and explaining our actions to foreign audiences. At this time this kind of all-out liaison and exchange of information and thinking and co-ordination of planning does not exist. The logic requiring it seems inescapable.

No longer are our military plans the concern solely of a general staff. They are governed, among other things, by our economic capabilities, our industrial potential, our technology, and a rapidly increasing knowledge in almost all fields. Similarly our political plans, our desires in relation to the actions we wish to see taken by foreign nations, are conditioned by the kind of force we may be able to bring to bear if our diplomacy fails. Both the Department of State and the Department of Defense know that neither can proceed independently of the other in the modern world. Our military officers are represented in the formulation of national policy, and the complexity and far-reaching influence of our political requirements may be measured by the military's recognition that in all probability the last major war has been fought in which a political adviser is not attached to a theater commander's staff.

If, in the conduct of our national life, we have been able to comprehend and act in conformity with the facts that have combined to bring about a measure of interdependence among powerful departments of government responsible for our welfare, it is only logical to assume that these same facts act to require a similar interrelationship when it comes to explaining and interpreting our actions. This, unfortunately, is not yet fully recognized. Or, perhaps more accurately, it is recognized, but as yet the obstacles in the way of successful liaison in the field of political communication have not been surmounted. In part this is because of what must appear as a deeply ingrained reluctance to turn to the instrument of political communication. If only the instrument were accepted in the public mind as the major departments of government are accepted, it is unlikely that the obstacles to successful liaison would persist. It is because, among other things, many honest, hard-working government officials are not convinced that political communication ranks with force and politics that the obstacles to effective liaison and co-ordination have continued to this time and still continue.

The liaison that is necessary, and is being discussed here, is not the conventional assignment of an individual or two from one service to keep perfunctorily in touch with the operations of another. This kind of liaison does, in fact, exist. What is required is active partnership and participation among the agencies concerned with the forms of political communication we are currently carrying on.

This means that USIA and the military establishment, for example, should each be fully aware of each other's plans and operations and that the daily activities of each should be shared in by the other. In the event of war the military establishment will have a need for a going psychological warfare organization. It is logical to assume that it is now planning for such a contingency. But what the military establishment does in the way of psychological warfare cannot escape taking into account our activities in the field of political communication prior to the outbreak of war.

Normally attitudes change slowly. If we have been trying to change certain attitudes before hostilities, it is likely that we shall wish to continue to do so. The work that has been already done will need to be reinforced, not cancelled and a fresh start made. This continuity of purpose cannot be effective without awareness of what has preceded a given date in time. The kind of liaison that will permit this calls for a team approach. Furthermore, it is virtually inevitable that members, at least, of the agency in operation before any war will be called into service to continue their work for the military establishment. If they are, they will need to understand military procedures, organization, and command channels. Again, the way to get this understanding is by active, shared liaison. The kinds of operation appropriate in peace and war, and their relationship to each other, need constant examination, discussion, and preparation. The most logical way for this to be accomplished is for members of both agencies to exchange personnel, share their plans, discuss their problems, and alter, modify, or conform their practices as a growing understanding of the roles of each becomes apparent to the other.

This is not the kind of liaison that is apparent today. Furthermore, if political communication is to be effective, those engaged in it must be aware of, and consulted and advised by, the appropriate officers of such departments as Defense and State on the contemplated courses of action our national policy requires. The liaison, again, must be in terms of a team. This kind of sympathetic understanding has not yet been achieved in spite of all the experience and the lessons of the past.

In totalitarian hands, both German and Russian, one of the facts that is clearly underlined is the necessity for the actions of government and the propaganda about those actions to be integral and co-ordinate. This is relatively simple for authoritarian governments to bring about. It is difficult for a free society. For one thing, a totalitarian state possesses "doctrine," a group of beliefs that it is quite willing to try to impose on others, by force if necessary. Diametrically opposed to this is our concept in the United States. We are chary of doctrine and leery of imposing beliefs on others. Doctrine, in fact, we disdain, if for no other reason than that developing facts and later knowledge so often act to upset it. Our history, and particularly our history of ideas, shows quite clearly that we tend to accept the facts available at a given period, but always with the reservation and the proviso that we will discard these facts if later knowledge shows them to be in error in certain particulars.

This obviously makes it difficult for us to speak with a united voice on issues of importance. It ought not to mean that we are estopped from developing an organization and the necessary team play among departments of government that make it possible for us to speak with emphasis and coherence on matters of national policy, whether we are at war or peace.

The Need for a National College

In order to bring this about, it may well be that what is needed is a college of political communication where the principles of the art will be taught, not where practitioners of the activity will be trained. It is only when the significance and value of political communication are thoroughly understood by the various agencies of government as a whole that we may expect the kind of team play that is suggested by the preceding discussion of the liaison requirements.

Veterans in this field are unanimous in their opinion that operations cannot be divided among departments of government with responsibility switching back from the Defense Department to the State Department depending on whether or not we are at war. But this recognition of the nature of the instrument is by no means so generally held among the ranks of those whose experience is limited or nonexistent.

If the social scientists who have studied the subject, and the practitioners of it, are right, a strong case can be made for the establishment of a college along the lines of the National War College to which accredited representatives of government may go.

Something approaching such a national course was essayed in the summer of 1950. It was discussed among various departments of government and at various levels for some six months. It was projected, however, more as a training program than a teaching college. When discussion of it was initiated, it had three aims: first, to acquaint government officials with the capabilities of political communication;[16] second, to train appropriate individuals in the

[16] The papers of the time, of course, talk of "psychological warfare," not political communication. The context, however, clearly shows that it was in terms of what this book calls political communication that the discussions proceeded.

essentials of planning and policy formulation for the activity; third, to make provision for a reserve, which could be activated in the event that a major psychological warfare operation was considered necessary.

At the outset it was an ambitious program that was conceived. The discussions were in terms of a thousand officials of the government who would take the training course. They would come from the military establishment and the Department of State. Later in the discussions the Central Intelligence Agency was included.

One of the main reasons for suggesting such a training program was that there existed no provision at the national level for adequate training of qualified individuals in the fields of foreign information and psychological warfare for the discharge of executive and policy-making functions.

As the discussions proceeded, one of the three aims originally set forth—namely, the training of practitioners—was dropped. The emphasis was centered on the executive aspects. A broad course of instruction was proposed. It included such subjects as the theory of international politics, the practices upon which United States policy-making depended, the way the United States government was organized for the conduct of its foreign policy, information, and psychological warfare operations, and case studies from our past history of planning and operations in the area of political communication.

Discussions on this ambitious program went on for some six months. At the end of that time the whole idea was quietly and effectively shelved. Virtually all of the departments and agencies of government that had been concerned with it decided finally they were less than enthusiastic. The reasons for negative action after an auspicious beginning are complex. But it is the writer's opinion that the evidence, when finally it may all be examined and studied, will show that one of the reasons was the difficulty of getting agreement on the proposition that political communication calls for a team and that this team is government itself. In other words, the importance of liaison and recognition that political communication is a national and not a departmental undertaking had not been clearly grasped.

This recognition is still not generally accepted today. It is difficult, however, to see any other possible conclusion after the nature, the premises, and the practice are closely examined.

A national college that would teach, not train, might do much to bring about wider and increasing acceptance of political communication as an instrument of state. Its faculty and courses should be such that it could develop as a great school. This means facilities for research in addition to teaching. It would require a permanent faculty of men willing to make the work of the college a career. Additional faculty should come from practicing officials in the departments of government, attached for periods of one or two years—men who were specialists in the field of political communication, its informational, economic, diplomatic, and military aspects. Finally, visiting lecturers, men eminent in the various fields of especial interest to political communication, should be invited for shorter periods ranging from six weeks to three months.

Considering the responsibilities of the United States and the likelihood of a prolonged period during which the psychological aspects of policy will be of strategic importance, a national college for teaching the fundamentals of political communication, possibly developed along the lines of the National War College, could be justified as necessary if the United States is to succeed in leading the free world.

It is a major premise of this book that at this moment in our history there is a great need for the application of knowledge and scientific method to the solution of the problems we face. This knowledge exists and is increasing rapidly enough to permit the conclusion that, if it is sought out and applied, we may well succeed in overcoming the difficult and dangerous years ahead without a general war.

One of the tools by which this may be brought about is a national college in which the problems we face may be rigorously examined by specialists in the various fields concerned with the minds of men. It is ideas that are rampant in the world today. New ideas in conflict with old, and old ideas and old traditions cloaking and hampering the discarding of outworn knowledge supplanted by later discoveries and increasing facts. More and more we are turning to the laboratory for proof of this knowledge, and mere authoritarianism cannot survive the proof of the experiment that can be repeated. It is the tripartite nature of man that is the key to the present struggle, and, if the spiritual ideals of free men are to win, they must be supported by knowledge, and this knowledge must be debated and tested. This is the role of a national college of political communication. If the reading of the evidence is correct, if, in fact, the instrument deserves and requires equality of status with our diplomatic, military, and economic activities in the field of international relations, then such a college is not too much to ask, nor conceited to plan for.

The Ideals and Objectives of the United States

It should be clear that the political communication of a free society must be instinct with the avowed and latent aspirations of its people. At the same time it must be supple and flexible enough to react with vigor and decision to the many voices that pretend to speak with authority in an epoch marked by tremendous fluidity of thought and great strides in knowledge. Those who are charged with responsibilities in this field need charity and tolerance at least as much as they need toughness and resiliency.

What we are seeing in the world today is a rapid advance in man's understanding of the laws that govern and condition his environment. At the same time the decisions that must be made rest on values that have come down to us from the beginnings of history, values that would be as clear and comprehensible to Akhnaton or Aristotle as the accomplishments of the sciences would smack of wizardry.

Loyalty, fortitude, justice, and prudence, for example, are virtues that still possess power to attract and influence our relationships. But we cannot quantify them. It is difficult to assign numbers or degrees to them. Yet they possess reality, if by reality we mean that to great numbers of men they are accepted as factors to be taken into account in making decisions.

Behind these concepts, in our own society, stands the Constitution, which both implicitly and explicitly accepts values as ends—liberty and the pursuit of happiness, for example, and freedom of belief.

It is not by accident that the United States today is the leading power in the free world, perhaps, even, the leading power throughout the world, challenged only by Russia. As Adlai Stevenson has said, the symbols of the beginnings of this power were the axe, the plough, and the Bible. The laws of the universe, in other words, are married to the values of the human spirit, and both are necessary if men are not to live in anarchy and chaos.

It is the nature of this relationship that has concerned mankind from time immemorial, and it is the thinking of men on the problems this relationship raises that gives us the ideals of a people and plays a part in the determination of national objectives.

Political communication is concerned as much with this aspect of human behavior as with the necessity for preserving the national entity, for it is the force of ideas, history shows, that has dominated the development of societies, equally as economic and political factors have done.

No people may escape its past in pressing forward to its future. It is in terms of its past that its future will be secured. And the past history of the United States demonstrates, among other things, the growth of a people who have been reluctant to see power concentrated in the hands of a few, who believe in the wisdom of majority rule, who have accepted as a principle the determination of issues by the people whom those issues affect and not by imposed judgment or oligarchic edict.

In our political communication to the world, the values on which our society rests must be as evident as the accuracy of the knowledge by which we exercise our power. In the light of knowledge, facts as we have understood and accepted them may change, which is only another way of saying that on many questions that perplex us, all the facts are not yet in. With increasing knowledge the courses of action we may elect at a given time can alter, and the change does not have to impair the moral value of the currency of our ideals and aspirations. But the principles on which as a nation we have been founded may not change without an attendant risk that our integrity becomes questionable.

The integrity of a nation is made up of its character and actions and the principles it proclaims. Knowledge may change its actions without hurt. But if, too, its character and principles change, its stability is in danger. It may no longer expect to lead, and it cannot effectively retain responsibility.

The objectives of the United States must be tuned to its ideals if they are to persuade and lead. The nature of those ideals is intangible, is rooted in values, and is discernible chiefly in the history of ideas. The art of great political communication lies in action that is consonant with the great ideas that have made us a great nation. It must be evident in the explanation and interpretation of our actions that we have not disregarded the link between knowledge and values. It is the truth of values as well as the truth of knowledge that gives to great ideas their force. Both must be rigorously tested and constantly re-assessed. It is this continuing search that gives mankind dignity, and it is the climate that permits this search that allows us to hope we can persuade and convince men of the worth of freedom when it is challenged by authoritarianism.

VIII The Problems To Be Overcome

If we are to develop and use the instrument of political communication to an extent commensurate with its potential capabilities, we must realize that a certain pattern has been created by our past history and experiences with the activity and that this pattern requires some alterations in essential particulars.

We have to remember that a considerable amount of our recent experience is rooted in war. Ever since the end of World War II it has been in terms of the cold war that we have resorted to political communication. But if the concept put forward in this book is accepted, political communication is neither solely nor chiefly military in context. It must be regarded, rather, as an integral instrument available to government at all times for the furtherance and attainment of national aims.

On the basis of the premises that govern it, political communication is an honorable and important means, ranking with the instruments of diplomacy, the military art, and economic measures, which can be effectively employed, not intermittently, but as an essential component of the daily practice of government. We are more accustomed to think of it, under the name in current use—psychological warfare—as a military weapon. In time of war this aspect of it assumes prominence. But at all other times political communication possesses a more significant and wider role as an instrument of government for the prevention of hostilities and the preservation of peace.

It is the capabilities and requirements of political communication as an instrument, and not its subordinate aspect as a weapon, that require development. The problem here is one of grasping the distinction between the part and the whole when, in the past, our experience has weighted the scales heavily on the side of its military characteristics.

We have to consider, also, our ingrained reluctance to attempt to influence the thinking of others, except in times of stress. This has been referred to as a business contrary to our ordinary practice. This comment reflects a generally accepted viewpoint. And, indeed, if political communication is thought of as chiefly broadcasting and propaganda leaflets, there is a good deal to be said for it. But political communication is not a matter of slanted or selected information. It is, in essence, as has been said earlier, the interpretation and explanation of a nation's action and policy. If this is accepted, there should be no reluctance to our making the fullest possible use of it. The attempt that is being made, in these circumstances, to influence the thinking of others is not in terms of trickery or only partial revelation of the facts. The objective is the presentation of our national position so lucidly and unequivocally that all who are affected will understand clearly what we mean and what we intend. And it should be our hope that these intentions will affect not only thinking, but decisions, in terms of relevant political action that will accord with our aims. Obviously our aims, in such circumstances, must be congruent with the basic desires and aspirations of free men. But this assumption is implicit in any program of political communication undertaken by the government of a free society.

Furthermore, we shall have to stop thinking of this activity as propaganda in an odious sense, and the concept must be broadened to include far more than merely information work. Information work, in fact, which is what is thought of now, for example, when the activities of USIA are considered, becomes of secondary concern. Instead of merchandising American culture, American literature, arts, crafts, and political practices, the task takes on the aspect of militant interpretation of the significance of the actions and decisions of the executive, legislative, and judicial branches of government. To the extent that we can demonstrate that these are consonant with the common desires of mankind, we may hope for acceptance and thinking that generally will support our moves. In the event that we are narrow and chauvinistic in our actions, we ought to expect quick rebuttal and evidence that should, if political communication is to fulfill its function, impel us to reconsider and re-assess our proposed courses of action.

Beyond this, we must also accept the instrument as auxiliary only, never primary. In the past, at times, there has been a disposition to look on it as capable of accomplishing ends by itself and on its own. The evidence for this is slight and the logic for it unconvincing. The minds of men are not changed by words—they are changed under the impact of dynamic ideas translated into concrete action that proves the conviction behind the idea.

In trying to accomplish our purposes by propaganda alone we can be shown largely to have failed, and for reasons previously given. This is a mark of immaturity. In the past, too often, we have been fired with missionary zeal for what we have termed the American way of life, and we have set out to try to convert others, perhaps almost as literally as the missionaries of the nineteenth century tried to convert the "heathen," and this is an impossible and thankless task in which we are vulnerable to the ancient admonition: physician, heal thyself.

We have to remember that both we and those to whom we may be talking still have great gaps in our knowledge and in our practices and that we know nothing that permits us to assume that we can get away with preaching. We can draw, and lead, and assume responsibilities, but we can do this successfully only if we demonstrate that we, too, are learning even while we seek action that accords with our objectives.

A further problem for attention centers on personnel. The history of our experience shows that we have had many unhappy moments involving personnel. In World War II psychological warfare was, essentially, a civilian activity carried on in theaters of military operations. Many of the qualifications for the kind of work that had to be done called for men who, in Sir Robert Bruce Lockhart's words, were prima donnas. The characteristics of military discipline and the antics of prima donnas do not marry well.

As a result of World War II experience, the Army wants its personnel in this field, in any future hostilities, to be in uniform. In the Korean war the Army ran our psychological warfare operations in the Far East. Relatively few civilians were involved, and those who were came under the classification of DAC's (Department of the Army Civilians), which gave the Army the authority and control it desired.

Recently the Army has been tapping men currently engaged in information work and arranging to use them if hostilities break out. This, in effect, is to weaken the national effort. The major pool of competent personnel for this kind of work is USIA. In the event of war, USIA ought to switch its emphasis from information to psychological warfare and carry on its duties. It will not be able to do this if many of its staff are drawn off into the Armed Services for military psychological warfare duties. And furthermore, strategic psychological warfare is not, under the basic premises, the prerogative solely of the military establishment. It is, first and foremost, national and political in character, not military. Our strategic psychological operations, in the event of war, will require the skilled services of men now in some phase of civilian life; many of these men are now serving USIA. It is these men who will be the backbone of any major psychological work we undertake in war. The problem of how best they may serve the national effort with their skills and competence is one that calls for careful consideration at high levels in the government. If the problem is not weighed now, and decisions taken, it will in all probability be too late, if war comes, to do anything more than make hurried *ad hoc* compromises. This is what has happened in the past. And the results have not always been fortunate.

At the same time that the problem of personnel is considered, the dilemma of security also calls for attention. This dilemma has its roots in the fact that political communication wants and needs all the intelligence it can get. Intelligence, traditionally, especially in wartime, is surrounded by an aura of secrecy. Political communication wants and needs secret information in order to talk openly to all sorts and conditions of men. This requirement is in diametric opposition to the need for secrecy in intelligence work.

In the past the interests of secrecy have usually been sufficiently powerful to prevent our information officials from having access to necessary knowledge. There can be no quarrel with the importance of secrecy on issues affecting the survival of the nation. Recognition of this fact, however, ought not to mean that senior officials responsible for our political communication may not have access to plans and policies.

This is a knotty and delicate problem. On its solution depends, in no small degree, whether or not we may expect to do effective work in this field.

Furthermore, the premises on which the whole structure

of a national program of political communication rests have not yet been thought out and accepted. For example, there are many who believe that it is important for a free society in work of this nature to base that work on the truth. But there are others who believe, equally honestly and without cynicism, that it is quite permissible to lie under given circumstances. This problem relates directly to the different roles of what are called white and black operations.

In this book an attempt has been made to examine and order the basic premises on which political communication may be undertaken by a free power. But there is, as yet, no general acceptance of them. There is, in fact, no codified arrangement of them. It has thus been possible, in the past, for the heads of separate but related operations to direct their staffs on the assumption of conflicting premises. Self-evidently this reduces over-all effectiveness.

Still another problem concerns the integration of this instrument into the complex of government machinery and departmental operations. We have now, for example, aspects of this work under the jurisdiction of the United States Information Agency, the Committee for Free Europe, the military establishment, and the Central Intelligence Agency. There ought, manifestly, to be co-ordination and liaison and central direction, both in terms of those agencies operating now in the context of a cold war, and in terms of those that will be operating if the cold war turns hot. There is not now this central direction.

To a considerable extent our activities in the field of political communication have been growing rather like Topsy—somewhat undirected and unplanned. Precedents have grown up or are in process of growing up. Unless steps are taken to correct and alter certain trends that are evident in the history of our operations in this field, we almost certainly face the prospect of not being able to make really effective use of this instrument.

Some of these precedents have been discussed briefly above. They will be discussed further in the next chapter when the question of how political communication may be organized and used is considered. In sum, a large part of our thinking about political communication has been in terms of war; we have accepted, without sufficient examination, the proposition that in going into this field we are tampering with other people's thinking, and this is repugnant. We have thought of operations in this field as involving chiefly what we call propaganda in an odious context; we have tended to regard it as separable among departments, depending on whether we are actively prosecuting hostilities or not; we have taken the position that decisions on whether the intelligence that information operators must have can be made available rests with intelligence authorities and not at higher levels.

These are among the problems to be overcome if we wish to use the instrument of political communication effectively and if we are to counter the enemy by it.

Intelligence

The problem of intelligence is sufficiently dynamic, important, and complicated to justify particular consideration in some detail. There are two aspects to this problem. The first is the actual provision of information—that is, intelligence—to those responsible for explaining and interpreting our official actions. The second is the concept of security—which, in effect, means keeping as much intelligence as possible in the fewest possible hands.

In World War II the available evidence seems to indicate rather conclusively that it was the concept of security that dominated the problem of the information agencies' obtaining the intelligence they needed. Military plans, certainly, and to some extent probably also policy, were drawn up and debated without the appropriate information officials being either consulted or advised.

This was not just a matter of military personnel refusing to provide important and necessary information. Some of the information personnel, in particular those who were serving overseas, were reluctant to fight the issue on its merits and on the logic of the position that unless there was access to plans, there could not be effective operations. Partly this was because, in the field, knowledge of plans was less difficult to come by than in Washington. There is a certain camaraderie once men get into the field. The atmosphere is different. Furthermore, the theater commander has an almost absolute control. It is difficult to get in without his approval. And it is difficult to get out without his concurrence. Once in, therefore, the atmosphere is rather reminiscent of a club. Information becomes available that it would be impossible to get in the bureaucratic atmosphere of Washington unless "channels" had been set up for the purpose. So, in World War II, it seems accurate to say that important information that should have been given to information officials was withheld repeatedly by military agencies. But it also seems accurate to add to this that information personnel in the field managed to get hold of what they wanted to know and were unwilling to risk their club membership by active participation in the continuing struggle at departmental levels in Washington to determine access as a right and not as a favor.

It was in the name of security, for example, that Office of War Information officials were denied information on the North African landings. Yet OWI participated in the operations. To complicate the problem further, OWI men in the field knew about the plans, yet they were not able to tell their superiors in Washington what they were up to. Nevertheless, they required supplies and men.

To some extent the intelligence the propaganda personnel needed so that they could plan properly was provided by their British counterpart numbers. British authorities had learned in World War I that information work could not be expected to succeed without its appropriate officials being given access to intelligence they had to have, if they were to plan effectively. In World War II this intelligence was given to the Political Warfare Executive (Britain's PWE, counterpart to our OWI and OSS). Some of this intelligence was passed along by the British to their American opposite numbers, for both OWI and PWE worked together in the European Theater of Operations. This is not the most desirable way for the members of what should be a team to get intelligence.

As the war progressed, the difficulties eased. But not because the problem of security was faced and mastered. Rather it was because both the United States military and civilian personnel working in the field of psychological warfare came in the course of time to understand better each other's viewpoints. This increase in mutual understanding, on what was essentially an informal basis, led to necessary information becoming less difficult to obtain.

Nevertheless when the war ended, the struggle for adequate intelligence, which had centered on the question of the right of access to military intelligence, had not been satisfactorily settled. One of the points at issue concerned whether or not civilians should be given intelligence of a sensitive nature. What is now called the need-to-know principle does not seem to have been working at that time, probably because, for one thing, the requirements of the instrument of political communication were only imperfectly comprehended. There was never, apparently, a ruling on this issue. At times, and on occasion, access was obtained, but the right of access was not established.

At the end of World War II and up to the outbreak of

the Korean war the issue of intelligence seems to have slipped somewhat into the background. Korea once again brought it into prominence. But this time there was a difference. Psychological warfare in World War II had been a mixed operation. Soldiers and civilians were combined and worked side by side. The Korean psychological warfare operation was a military affair from start to finish. A leading officer on General MacArthur's staff is popularly reported to have observed, when psychological warfare got under way: "There will be no goddam civilians in this operation." In the sense that there was no civilian *authority*, as, for example, with OWI in the field during World War II, he proved correct.

Interestingly enough their absence did not solve the problem, but it did alter it. Whereas security had been the troublesome aspect in World War II, in the Korean war it became the actual provision of intelligence. In other words, the Army was quite willing to give its own officers access to intelligence. But then it turned out that there was considerable difficulty in knowing what intelligence psychological warfare required.

Both problems are still with us. It is probably safe to prophesy that in any future war soldiers in uniform carrying out a psychological warfare mission will get access to intelligence sources. If they are civilians, there are probably going to be difficulties. In fact, unless some decisions are made, it is virtually certain that there will be difficulties. At the same time, however, the problem of access is complicated because political communication, as yet, cannot specify with precision the kinds of intelligence it needs and wants.

Some of its needs, of course, are known, but there is no thorough answer to the question: What type of intelligence operation should be set up for political communication? Opinions vary on the nature of the requirements of intelligence for this activity, on the sorts of training necessary for intelligence personnel in this field, on the relation of intelligence for political communication to national intelligence, and on the relation of the military aspects of this instrument to military intelligence.

Yet there is general agreement that intelligence is close to the heart of political communication. In fact the success of psychological warfare operations depends on highly specialized kinds of intelligence.

Colonel Kehm, who, during World War II, was the War Department representative of the Joint Chiefs of Staff liaison officers with the overseas branch of OWI, has written:

> The intelligence requirements for psychological warfare operations are enormous. . . . The political propaganda agencies can give the best support to military psychological warfare only when they have lots of material normally procurable only through military intelligence. Among the kinds of items needed are: recordings of battle and behind-the-lines scenes; talks by prisoners, refugees and escapees; important speeches, orders and proclamations by key commanders; communiques; captured reports, letters, orders and other documents; descriptions of liberations policies and procedures and surveys of reactions to our own forces and to military government. It is also essential that routine order-of-battle information be supplied and supplemented by information on enemy personalities and units. Obviously, no intelligence stone bearing on the status of enemy morale can be left unturned.[1]

Dr. Charles A. H. Thomson, who served in the Military Intelligence Service of the War Department General Staff, first with psychological warfare operations in Europe and later with military government in Germany, has two illumi-

[1] Col. H. D. Kehm, USA, "The Methods and Functions of Military Psychological Warfare," *Military Review*, XXVI, No. 10 (January, 1947), pp. 12-13. Command and Staff College, Fort Leavenworth, Kansas.

nating comments. His first is related to the continuing problem of intelligence.

> Wartime experience revealed that the detailed demands of propaganda on existing intelligence agencies far transcended their resources. Even for the current program, there is a continuing demand for basic political, economic, social, and military information, and for a wide range of detailed information about problems and personalities in foreign countries, both for policy and operating purposes.[2]

Dr. Thomson's second pertinent observation describes what intelligence in this field is required to do.

> It is the broad function of propaganda intelligence to do three things: (1) to provide a complete set of evaluated facts about conditions affecting propaganda in all countries abroad in which our information service is operating; (2) to provide similarly complete information about the social structure of these countries, so that propaganda targets can be chosen intelligently; and (3) to provide complete information about the general effectiveness of our propaganda techniques and operations, with considerable specific information about the impact of particular programs upon their recipients. This last sort of information is of particular importance in demonstrating how propaganda actually accomplishes its portion of national strategic tasks. It is the kind of information which policy makers, operators, and those congressmen or executives responsible for judging and controlling the program should have as the foundation of their work and of their criticisms. This is the sort of information which intelligence has not been able to provide, with the staff available.[3]

Since these writings, a greater flow of intelligence for the work of political communication has been provided. But it is chiefly an increased flow, not more precise intelligence,

[2] *Overseas Information Service of the United States Government* (Washington, D. C.: The Brookings Institution, December, 1948), p. 338.

[3] *Ibid.*, pp. 344-45.

nor intelligence tailored to the instrument on a foundation of exact knowledge. There are good grounds for holding the opinion that much more could be done to make intelligence for the activity both more responsible and more effective.

The Inertia of Public Opinion

Public opinion in a free society is one of the controlling factors in determining whether or not political communication is able to do its work effectively. As long as political communication is thought of as propaganda in an odious sense and equated with it, the grounds seem good for defending the position that it is not likely to be worth much time, money, or attention. If the United States is to match Russia in this field, there must be a radical alteration in the general opinion about the value of the instrument.

This alteration obviously cannot be accomplished by fiat. It is the record of results that counts. The apathy of public opinion is adequately attested by its indifference to the repeated assaults that have been made on the program of the United States Information Agency. In part this must surely be because it is difficult for USIA to show conclusively that its work has been worth while. As Barrett says, "The accuracy of a gun or the speed of an airplane is subject to clear proof. The effectiveness of persuasion rarely is."[4] The evidence that political communication is effective is not "hard" evidence. Furthermore, it requires considerable time, reflection, and judgment to put together the nature of the case on which the argument for political communication must rest. In the accepted sense of the term, this does not constitute "clear proof," although, if it is done, it assuredly seems to show strongly that the activity is a powerful one.

[4] Edward W. Barrett, *Truth Is Our Weapon* (New York: Funk & Wagnalls Co., 1953), p. 255.

It is in terms of tedious and protracted examination of the record of what we have called psychological warfare that this book takes the stand that political communication potentially has powerful capabilities that can materially help us to attain our national aims and objectives.

This kind of examination, obviously, will never be made by the general public. Yet the support of public opinion is one of the essentials in political communication. Not until government shows that it really possesses confidence in the instrument will public opinion alter. The requirement, then, is not that the general public be asked to weigh the factors leading to the conclusion that a free society should employ political communication, but that appropriate government officials should do so, and, if they find, as this writer does, that the conclusion is justified, the necessary steps should be taken to translate the conclusion into positive action.

In other words, public opinion will not, of itself, be able to give political communication the kind of support it must have if it is to be effective. There must be leadership. It is worth remembering Frederick C. Irion's words in this connection. "Public opinion, more than any other force, will determine how people think and act,"[5] but, at the same time, we must also remember that

> Because public opinion in the United States is popularly regarded as a moral force, one note of caution regarding its influence should be sounded here. Public opinion as a social phenomenon *per se* has nothing within it making it inherently a power either for good or for evil; without direction it is amoral. If it is to serve as an aid to democracy, this result can accrue only through the determination and efforts of people-of-good-will; otherwise the possibility of enslavement of people through public opinion, directed by propaganda, is always present.[6]

[5] *Public Opinion and Propaganda* (New York: Thomas Y. Crowell Company, 1950), p. 758.

[6] *Ibid.*, Preface, p. vii.

The leadership, then, that public opinion must have raises, in turn, some delicate and difficult problems. In corrupt or evil hands public opinion could drive a nation to its own destruction. Perhaps as good an example of this as exists is Germany under the leadership of Hitler and his disciple Goebbels. It is quite conceivable that, through mistaken understanding of the power of this force, our own people could be led into a position where, from lack of knowledge, or only partial knowledge, public opinion would support a course of action demonstrably false if all the factors necessary to wise decision were available. In Irion's words, again, "As demonstrated by World War II, it is easy to direct American thinking along one path if there is sufficient emergency. However, the managers of public opinion during World War II were amateurs compared with what might be achieved."[7]

It is probably by reason of an instinctive grasp of the truth of this that Congress so consistently has objected militantly to any United States government agency for propaganda directing that propaganda inward to its own people. We face here a dichotomy in our thinking. Oversimplified, it is wrong for a government agency to attempt to influence the thinking of Americans. It is permissible to attempt to do this to foreigners. It is this conflict that is one of the reasons why what we have called psychological warfare requires rethinking, and why the concept of political communication ought to supplant it. If, as a free people, it is wrong to tamper with the right of our own people to form their opinions, a good case can be made for taking the position that it is wrong to try it with others.

Political communication ought not, if the basic premises are accepted, to try to tamper with thoughts and attitudes. It is based on facts, on reaching conclusions supported by

[7] *Ibid.*, p. 755.

knowledge—knowledge that can be tested by all men and should be available to any who search for answers.

No one country and no one people have a lien on knowledge. Its sum is made up of the contributions of all nations and all peoples. A two-way process, which is what communication implies, therefore must mean that the principle of learning is the principle governing our use of this instrument in the international field.

It is the view of this writer that the public is unlearned in what is meant by political communication and unversed in its capabilities, but that it would be prepared to support political communication if it saw government giving it weight and standing and if government made a determined attempt to lead and inform the public about what the activity in reality means.

One reason for public apathy in this, as in other, fields is privatization.

> Privatization is the awareness of the many that they can neither understand nor affect to any great extent the social system in which they live. "The common man feels impotent in a world where specialized skills control events that at any moment may transform his life. . . . Two attitudes result from this feeling: first, one does not inquire into the causation of the events thus viewed; . . ."[8]

In the field of political communication, the "causation" of many of its activities can be traced ultimately to public opinion. If this opinion is active and informed, we may expect positive courses. If it is apathetic and listless, our political communication will be weak. Leadership of public opinion is essential, not in terms of trying to tell the public what it should think, but in terms of giving it the necessary information on which to come to a conclusion that, once it

[8] *Ibid.*, pp. 659-60. Inner quote: Ernst Kris and Nathan Leites, "Trends in Twentieth Century Propaganda," in Geza Roheim (ed.), *Psychoanalysis and the Social Sciences* (New York: International Universities Press, 1947), p. 401.

has been reached, will serve to guide those responsible for political communication in deciding how best to interpret and explain the actions of the government.

An informed and virile public opinion in the United States could have far-reaching consequences on the ultimate issue of today's events, but, in a constructive sense, only if it is given adequate information and the nature of its role is made clear to it. The steel strike of 1920 serves as an example.

> In sum this strike is an illustration of an extreme use of public opinion by an industry. It also shows the force of public opinion. When misinformed, the public was so hostile to the steel workers that their cause was temporarily lost; but when informed of the truth, the public and the unions forced one of the most powerful industries to change.[9]

The Dynamic of Ideas

When one looks at Soviet political communication and compares it with our own, what stands out is the completeness with which the instrument has been integrated into all forms of Soviet activity. "Communist propaganda aims at the total mobilization of all energies in man: spiritual, material and intellectual. . . ."[10] If we are to reply merely with counterpropaganda, or even propaganda, in the sense that the term carries today, we are at a disadvantage from the start. A free people cannot be so regimented. Freedom to disagree is basic to our form of society. This is another reason why it is so necessary to examine the premises on which a democratic government may undertake political commu-

9 *Ibid.*, p. 366.

10 Jean-Paul David (Deputy to the French National Assembly), "How to Win Countries and Influence People," *United Nations World*, October, 1951, p. 18.

nication, and it is why the premise that facts and knowledge should be one of the foundations of the activity in our hands is important. It ought not to be possible for us to support the sort of activity in this field that an authoritarian government undertakes. But, equally, this ought not to mean that we are unable to challenge it effectively.

The challenge, however, must be in terms of our own values, our own ideas, our own convictions. As Dewey has said, "All conduct that is not simply either blindly impulsive or mechanically routine seems to involve valuation."[11] At the root of these valuations there must be ideas. It is the strength and tenacity with which they are held, and tested, and revalued in the light of increasing knowledge, that give them force and validity. We, as yet, do not seem to have found the key that will unlock our energies and permit them to be purposefully directed against the insidious and demonstrably false ideas with which we are in conflict. As the Hickenlooper Report observed, "The short history of the American program [of overseas information] has been characterized by a search for means which would advance both the interests of the United States and the community of interests of the non-communist world."[12] This search is still going on. To repeat Lasswell's words, "We are in a war of ideas, but we have not found our ideas."[13]

This is not to say the ideas do not exist, but no one has appeared on the horizon who is able to express them in a form that will fire us with purpose and direction. In World War

[11] John Dewey, "Theory of Valuation," *International Encyclopedia of Unified Science,* Vol. II, No. 4 (Chicago: University of Chicago Press, 1939), p. 3.

[12] Report of the Committee on Foreign Relations, Overseas Information Programs (Hickenlooper Report), Senate Report No. 406, Eighty-third Congress, First Session (1953), p. 2.

[13] Harold D. Lasswell, "Policy and the Intelligence Function," in Daniel Lerner (ed.), *Propaganda in War and Crisis* (New York: George W. Stewart, Inc., 1951), Ch. 4, p. 67.

II Sir Winston Churchill was able to do this for his people, but the threat and cohesive force of actual war are not operating today. The nature of the present threat does not seem so immediate. It is an ominous shadow in the distance, but it has not actually darkened our daily life—yet. Its nature, however, is such that if it does, it threatens to destroy us, as no threat before has done. It is precisely at such a time that we need to be clear about the convictions and values that give to our way of life the qualities we feel ought to be preserved. Who is to state them for us in terms applicable to the mid-twentieth century and the decisions we must make?

This is, perhaps, our greatest weakness in the whole field of political communication—the lack of an expressed dynamic capable of knitting us together as we were knit by the fateful attack on Pearl Harbor.

In the absence of this spiritual force, political communication faces difficulties. The essence of the activity is the explanation and interpretation of the actions of government. But if those actions are chiefly reactive, political communication can be no more than reactive. We are caught here with an instrument that gives indications of considerable power for the attainment of our national objectives, but since it is an instrument that is directed to the minds of men, it is required to carry out its work in terms of the intangible forces that fire and capture the imaginations of men. "It [the science of propaganda] arose as it were at the cross-roads of other and more established sciences . . . In plain terms, there met in it the three sciences of mechanics, of psychology, and of ethics . . . it is clear that the propagandist can take scarcely a step which does not involve a reference to ethics."[14]

It is ethics that determines whether a people and a na-

[14] George Peel, "Science of Propaganda," *The Contemporary Review* (London), November, 1944.

tion act from principle or expediency. And, in the great struggle that now involves us, the people of the world are looking to us as the leaders of the Western heritage to see what we propose to do.[15]

Political communication cannot make the necessary decisions. It can only explain and interpret them. And it cannot interpret what is not there to interpret in the first place, nor in a free society may it interpret what cannot be explained.

Until we can find "the common purposes, the common interests, the common values of the free world,"[16] and express them in terms capable of firing the hearts of men and then prove by action that these common denominators are the mainspring of our endeavor, our political communication is unlikely to be more than counterpropaganda. As counterpropaganda the struggle must take on the character of the survival of our nation against the Soviet. This will not be the victory of the heritage of the Western world in conflict with forces denying the dignity of men.

The Political System

It is a basic assumption in this book that we are face to face not only with the problem of survival—which is important enough—but, in that survival, with the preservation

15 Note Charles Malik, of Lebanon: "Politically the West will not serve the cause of peace by allying itself with dark regimes just because it is more expedient not to disturb them. . . . Their peoples are poised to see whether the West acts from principle or expediency. And the subversive whispers of world revolution become more and more potent the more these peoples despair of their rulers and the West." "War and Peace," a statement made before the Political Committee of the General Assembly of the United Nations, November 23, 1949, The National Committee for Free Europe, Inc., New York, pp. 30-31.

16 George E. Taylor, "The Intellectual Climate of Asia," *The Yale Review,* XLII, No. 2 (December, 1952), p. 194.

of what is called the Western heritage. To survive by means that cost us this heritage is to postulate such a transformation of our national entity as to suggest a complete rupture with our past and a beginning again on terms difficult, perhaps, to imagine but assuredly unrelated to our present way of life. To inject this consideration is to point up the range and scope covered by the concept of political communication, for, in any examination of the instrument, it is evident that a peculiar admixture of qualities is demanded (*juxtaposition* is perhaps a better word than *admixture*).

On the one hand, substantive problems of a pragmatic nature call for analysis. At the same time, there is a requirement for controlled, disciplined idealism. This is because, as George Peel has said, the science of ethics is an indispensable element in the political communication of a free society. If we are unwilling, as a nation, to admit this, we had better be prepared to abandon any idea of a national operation in this field.

Some of the preceding sections of this book have been couched in "idealistic" terms for this reason. We must return here to consideration of a major problem, substantive in content, which is of direct concern. It is the problem of the machinery of government by which things get done.

To this writer it seems that the more an individual learns of the processes of our government, the more difficult it becomes to separate and explain in simple terms the strands that, taken together, make up the fabric of our political life and to assess the role they play, and the degree of importance they command, both in reaching decisions and in carrying out those decisions once they become policy.

It is absorbingly interesting, for example, to read, as this writer has, a directive prepared for high levels setting forth national policy—a directive comprising perhaps only a handful of pages in which both clarity and equivocation may be subtly blended—and then, with this end product as the starting point, go back step by step through the maze of papers from various departments of government that show how the final position was reached, the time this took, and the men who did the spade work.

It is by no means the men at the top levels whose thinking is always reflected, and it is by no means these men who give substance to the sometimes conflicting departmental positions with which the general public is familiar in the public press.

A careful reader of such documentation slowly becomes aware that there is a small army of middle men in government possessed of average qualifications, even as you and I, who spend long years in a departmental field while the arteries of mental flexibility harden under frustrating bureaucratic procedure, as they accumulate knowledge of how their counterparts in departments they work with think and react. Inevitably the give and take of compromise develops into a sort of game in which one line of a proposed document intended to set forth policy is exchanged for another, in which one paragraph, written by the members of one agency, clear and succinct in its first draft, is skillfully and often charmingly "fuzzed up" in successive drafts by members of another agency who see, in the original, a timeworn, sacrosanct position eroded if it is left untouched.

It is both funny and heart-breaking to read the successive drafts of such a document—perhaps twenty or thirty of them, stretching out over several months—and see how, in the final form, the art of "fuzzing up" has been employed. It is disconcerting and, at first, startling to discover that an original operational instruction at one level in a department, which is breath-takingly clear and simple to grasp, has been reworked so that limits of responsibility are vague and to learn that this may have been done to avoid the onus of blame in a delicate political situation.

Along with the successive drafts go series of conferences, which sometimes make little headway, so that a day's work finds the emerging policy only a sentence or two nearer its final form. For one thing, the blinders of established departmental positions blur the logic calling for initiative and original thinking to meet the dynamic and unstable events of the times. For another, the words spoken across the table often carry one meaning to a speaker and another to a listener, and the unspoken premises and assumptions behind the thinking of each are obscured and need time and care and skill to get at.

Most of the people who take part in these draftings and conferences are aware of what is meant by the phrase "talking past" each other. Relatively few have the gift of being able to listen and then translate into general comprehension the points on which there is agreement and clarify those that still need to be threshed out.

It is this sort of tangle that may well lurk behind the final numbered papers that are acted on and approved at the high levels of government. And the men in these high levels are not always versed in the history of the problem with which they have to deal. They have the authority to act, but they lack experience, for they are in political posts that change with administrations, while the men who have the knowledge, and serve those at high levels as they come and go, lack the authority.

Thus slowly, through trial and error, a climate is created, and in due time portions of public opinion become aware of this climate, and then advances can be made.

This is the background of our record with what we have called psychological warfare, so that now, as a result of empirical knowledge, there is a converging of informed opinion, which, if effective leadership is forthcoming, could result in major decisions permitting the instrument of political communication to function more effectively.

The difficulties of this nature involved in formulating a policy, however, are matched in the operational realm when it comes to members of two or more departments in the field having to work together to carry policy decisions into effect. Inevitably differences of interpretation arise, and usually these differences start far down the line between two conscientious individuals in allied, but separate, agencies working together under a section chief. He may be unable to resolve the differences because his authority is not sufficient. He may be unwilling to take the problem up the next rung of the bureaucratic ladder to his immediate supervisor, or he may take it up and be told that it is impolitic to lay it before the appropriate official for decision at such a time. There is a marked aversion to winning the reputation of being difficult to get along with. The difference smolders, resentment increases, friction grows. Finally it assumes proportions requiring action at high levels. By this time it has become a problem affecting the prestige of the departments concerned, and before it is settled a senior and experienced official in one agency or the other may be recalled and replaced as the price of the initial postponements.

All of this is familiar to the government worker. It is not as familiar to the heads of powerful departments who come and go. Yet if political communication is to function, both recognition of the problem and machinery to prevent its exacting so heavy a toll are essential.

If we pause to reflect on the use the Soviets have made of the instrument of political communication, we cannot fail to be impressed by the efficient and workmanlike manner in which they have integrated it into their system. This does not mean that a free people can adopt a similar course, but it does seem to indicate, first, that political communication can be effective, and second, that it is, inherently, amoral.

The thinking that governs its application in Russian hands is repugnant to us, but undeniably it has been successful.

In the years immediately following the close of World War II, for example, the Soviet Union brought under its domination, without fighting, nearly half a million square miles of territory and a population of more than seventy million people. To no small extent, the instrument that played a significant part in this was political communication. We ought not to follow the same pattern in our use of it, but surely we ought to devise a means for such an instrument that allows us to use it on terms that do not offend our ideals and aspirations.

In Russian hands political communication has been both flexible and responsive to initiative. Time after time in the past decade the speed and efficiency with which the Soviets have moved from a military posture to political action and then to economic pressure have caught us unprepared with any reply except one devised as the result of their challenge. The central idea behind the Russian moves has been extended in terms of all their resources, capabilities, and their whole machinery of government.

If we are effectively not only to counter but to assume the initiative, all the indications point to the necessity for our drawing on our own vast resources and mobilizing them. This is not only a matter of material mobilization. The task calls for spiritual skills and talents as well as industrial abilities. It is Robert Sherwood who has observed, in the early days of World War II, that "Spiritual preparedness for coping with powerful evil was required before it began to occur to people that some tanks and bombers and aircraft carriers might also be helpful."[17] The point is that guns and tanks alone are not enough. Behind the material mobilization of a nation's resources stand the ideas that govern the manner of their use. This is the spiritual force that drives a people. In Russian hands it is corrupt, denying the dignity of man. It

[17] *Roosevelt and Hopkins: An Intimate History* (New York: Harper & Brothers, 1948), p. xiii.

is not enough merely to be against this, for demonstrably the corrupt idea has power to attract the minds of men. The spiritual force that must govern our own actions and reactions calls for militant expression and machinery that will make it evident to all observers that there is a national will in the United States and that this will is being led and directed toward specific ends on which others may count.

To do this, one of the requirements is machinery for the resolution of conflicting departmental viewpoints and positions within our own society, together with the substitution of an approach that more nearly resembles a team of skilled players harmoniously determined to achieve a goal that has been set. Policy, once formulated, requires methods of measurement to ensure that it is being carried out. The departments of government need policing, in effect, in order that the executive leadership of the country can be sure progress toward the policy objective is being made. We cannot afford unresolved positions on important national issues because the resolution will tread on the toes of one department or another. On policy of moment a *national* position is called for, and in evaluating the degree to which it is being attained, independent judges, outside the departments of government, are required.

Colonel W. R. Kintner suggests that what is called for are the characteristics of competent staff work applied directly to specific operations.

> We believe that the present structure is not adequate. This book has described many and serious shortcomings in the organization and operations of the Defense Department. Although reasonable men may well discount some of these feelings or disagree with certain of the recommendations which we have advanced, we do not believe that our over-all judgment can be lightly set aside. We are convinced that an effective American response to the Communist challenge must include major improvements in our defense establishment. We have argued that

> such changes can and must be accomplished within the framework of our basic political institutions and beliefs, for we have no desire to undermine from within the political system which we seek to defend. But—and we repeat—ours or similar changes are essential—as a minimum.[18]

In drawing on our entire resources for the conflict in which we are engaged we should not overlook the contributions that can be made by private organizations and private individuals, as well as by government. The work of such organizations as the Ford Foundation, for example, could be drawn into our national effort and expected to have important influence in making clear to the world the strength of our purposes.

Such a foundation undoubtedly would resent any move that appeared to put it in the light of becoming an adjunct of government. It would not require, and should not be given, any such approach. More likely what is called for is the furnishing of intelligence to such an organization so that in making its decisions, and following its own policies and objectives, it would have the benefit of the widest possible knowledge of where we are succeeding, where we are barely holding our own, where we may be losing ground in what is a titanic struggle to hold and win to our side the allegiance and support of international opinion. By the same token, the views and informed judgments of the officers of such organizations ought to be taken into account in any mobilization of our national resources.

The Ford Foundation, for example, for the period October, 1951, through September, 1956, had set aside for its Overseas Development Program, to finance grants and projects, a total of more than thirty-eight million dollars. At the close of its fiscal year September 30, 1955 - September

18 William R. Kintner, in association with Joseph I. Coffey and Raymond J. Albright, *Forging A New Sword* (New York: Harper & Brothers, 1958), pp. 237-38.

30, 1956, it had allotted more than sixteen million for International Understanding.[19] It is absurd to think that its officers could not contribute valuable and powerful support to national aims and objectives in what clearly must be a concerted effort if we are to win. And it is in the realm of political communication that such private organizations and individuals could be expected to find their greatest service. For the diplomatic, military, and economic actions of government, obviously the administration of the day must assume responsibility. But the role of political communication, in terms of the basic premises on which it rests in a free society,[20] could well be greatly strengthened by the informed activities of private enterprise and without losing independence of action and decision in its own fields.

Only, it is suggested, in some such way can the concept of "privatization" be lessened among our own people, and only if it is lessened, and more and more people come to feel that in truth they both do understand and can affect the tremendous conflict that is the central fact of our times, shall we be able to mobilize the full range of our initiative and resources.

The precise nature of any solution to the problem of an organization spiritually prepared to cope, in Sherwood's phrase, "with powerful evil" requires examination and analysis. It can take several forms. But the requirement for a decision that is designed to cope "with the revolutionary challenges posed by the Soviet concept of total war and the accelerating surge of new technology" rather than "with problems of the past"[21] has surely been amply demonstrated.

Any decision must take account of the part played by Congress and public opinion, as well as provide the machinery that will give to the President the knowledge he

19 The Ford Foundation Annual Report, 1956, pp. 108, 184.

20 See pp. 61 ff.

21 William R. Kintner, *op. cit.*, pp. 10-11.

needs to act effectively as a national leader. The indications point strongly to the President as the one to act under our political institutions as these have developed, for only the President is accepted as speaking on critical issues in the national interest and above departmental conflicts. Unless some alterations are made, and made soon, in the pattern of our present decision-making processes, we may well find ourselves continuing to run second in the race that is now being run for the prize of leadership in a world that is being revolutionized by the swift advances of knowledge.

The outcome, if it is not to be disaster, or surrender on our part, requires that the minds of men be led, not forced, to an acceptance of the facts revealed by open testing and scrutiny of truth as knowledge makes it available. Political communication, in terms of its basic premises, has a major role in the success or failure of this mission.

IX

Forging the Instrument

We are in a war of ideas. If there are any who doubt this, let them contemplate the effects, at this point in history, of a depression such as that which marked the close of the twenties. Above the economic and political results assuredly would tower, in the eyes of the world, the idea that Western democracy was unsuitable for the government of men in a nuclear age. And, as our allies faltered from inability to keep their political and economic societies functioning in such a crisis, and as the neutrals pondered their position between the two colossi of the United States and the U.S.S.R., the oft-repeated Soviet assertion that our way of life contained its own seeds of destruction would gain seeming credibility. The idea that the Western heritage could lead mankind into a more stable world, marked by increasing material well-being and greater philosophical assurance, would be dealt a body blow. In such circumstances "even liberty must, if necessary, give way to the dictates of Security because, to be more safe, men are willing 'to run the risk of being less free.' "[1]

Such a development would seem to provide support for

[1] Edward Mead Earle, "Adam Smith, Alexander Hamilton, Friedrich List: The Economic Foundations of Military Power," *Makers of Modern*

the Soviet ideology, and it is reasonable to suppose that ideology would find much greater acceptance. In such circumstances we should likely find ourselves standing very much alone, our backs to the wall, our own survival in jeopardy, and Armageddon as our only recourse from surrender.

It is, surely, an idea with which we are in conflict at a momentous period of history. The significant point is that we may not oppose the ideology confronting us by another ideology for one simple reason: We have no ideology. The Soviets have demonstrated quite clearly that they are willing to impose their ideology and doctrine upon mankind if they can. It is the essence of the Western heritage that it is unable to do this, even if it felt it had sufficient knowledge. At the root of our developing society is an instinctive realization that our values and convictions and material success stem from the application of increasing knowledge as the problems we face demand changing decisions.

The expression of such a position in terms that can be grasped and comprehended by a whole society calls for genius, but the recognition that the position is there to be expressed belongs to many of us. We are aware that the nuclear age holds great promise—and scope for great disaster. We have not yet found how to express the hopes and fears of this age in terms that will link free men everywhere in a common effort to disprove the ideology confronting us.

What we are looking for has been well expressed by Peter Drucker.

> We need a discipline . . . that explains events and phenomena in terms of their direction and future state rather than in terms of cause—a "calculus of potential," you

Strategy (Princeton, N. J.: Princeton University Press, 1952), Ch. VI, p. 118.

Inner quote from *The Federalist,* 1787, No. VIII (New York: Modern Library Edition, 1937), p. 42.

> might say, rather than one of "probability." We need a philosophy of purpose; a logic of quality, and ways of measuring qualitative change; and a methodology of potential and opportunity, of "turning points" and "cultural factors," of risk and uncertainty, of constraints and variations, "jump" and continuity. We need a dialectic of polarity, one in which unity and diversity are defined as simultaneous and necessary poles of the same essence.[2]

It is the task of the philosopher to formulate this discipline. It is the task of political genius to express it in simple terms, common to the aspirations of men. It is the task of political communication to explain and interpret the actions of a free government in terms of such a discipline, terms that are common to all men.

This must be the aim behind any political communication undertaken by a free society. Meanwhile, in the absence of such a discipline, and while we are searching for it, the instrument must be organized to interpret the actions of a government made up of men in whom, instinctively, there is present this conviction of the capability of knowledge to give us the answers we seek.

This is a tall order, for it imposes upon us the task of proving the fallacy in the Soviet ideology while at the same time proclaiming by our actions that we are not seeking to substitute one ideology for another, but rather to advance, as knowledge makes available more and more of the facts that permit us to reach decisions, not on the basis of fiat, but because they are true and can be shown to be true. If true, then mere authoritarianism crumbles before them.

This is no easy or short-range objective. It commits us to the difficulty of struggling to preserve a free society, when all the facts that are implicit in its way of life are not yet susceptible of proof, against an adversary who is strait-

2 "The New Philosophy Comes to Life," *Harper's Magazine,* August, 1957, p. 40.

jacketed by no such scruples, an adversary who is well content to blend fiat with partial truths in the sure knowledge that the history of ideas shows that the specious claim can often hold its own sufficiently well to win powerful and corrupt adherents.

If this brief summary of the argument is accepted, if it is agreed that we are in a war of ideas, then it is not the established instruments of government that will pull us through, necessary as those instruments are, and vital to the final outcome. It is not solely our political sagacity, our economic skills, our military strength, that will sustain us. To these must be joined a militant conviction of the validity of our ideas, and our political, economic, and military policies must be formulated in terms of those ideas and explained and interpreted in their light.

An immediate requirement is the decision as to whether the instrument for this is psychological warfare or political communication. It is the position of this book that it is political communication and that a study of the recent history of what we have called psychological warfare substantiates this.

For one thing, if we elect psychological warfare, we cannot expect to match the Russians—at least not unless we change our national premises. It is unlikely that public opinion in this country would support an undertaking that regarded truth and facts as expediencies, to be used when they supported our side and discarded when they did not. To do this would strike deep into the heart of the position on which all our great universities are founded. The men who would be willing to direct and administer such an operation would not be able to join in companionship with men brought up in the traditions of the laboratory and the rules of reason.

Those who are familiar with the history of recent developments in our national capital will remember the intensity of the debate, only a few short years ago, on whether or not a free people could elect the advantage of being able to initiate a preventive war, swiftly and with its attendant requirement of complete surprise. The advocates of such a course lost their case decisively. It is the considered policy of this country to accept the disadvantage of waiting for militant aggression to take place before replying to it. The decision was reached on moral, as well as other grounds. Our knowledge, at the time of debate, did not permit the conclusion that we could take such action and at the same time preserve the principles on which we are founded.

The decision is not final, of course. And herein is the measure of our flexibility and the proof of an earlier comment to the effect that we have no ideology to peddle. We are free to re-examine and draw a different conclusion if later knowledge dictates it. But, standing on the national premises, it will have to be in consonance with them or we shall proclaim that they have changed. The Russians are bound by no such scruples. It is not necessary for them to state the reasons impelling them to fight, nor, in the mind of the public, is there any ground for thinking there are certain actions they will not take, save on the ground of expediency.

Were we to elect to wage psychological warfare against such an enemy, we would obviously labor under a disadvantage from the outset so long as our national premises remained. As long as these have force, it is political communication that is the instrument to which we must turn, not psychological warfare.

In electing to use the instrument of political communication, however, we are up against a natural and perfectly healthy obstacle, if the premise that the support of public opinion is necessary is correct. This is the patent interest and concern of Americans with the great opportunities our technology is providing for the increasing development of our own country. We are sufficiently large, both in geographical area and population, to be able to devote our energies and resources to the improvement of our own society, our own productiveness, our own comfort, without as yet facing the prospect of running out of jobs for everyone or finding that, if we are to keep busy, we need to go beyond our own borders. There are highways to be built to carry an increasing commerce from one coast to the other. With a population increasing at the rate of some three million annually, the demand is in step with industrial capabilities. There is before us the vision of nuclear power with its indication that, properly developed, we are about to enter an era of unparalleled opportunity in which we shall have both more leisure to pursue our individual pleasures and greater opportunities for work and the wealth that will enable us to enjoy such pleasures.

We see this and we know that there is much to be done to make the dream a reality. What is difficult is to grasp the importance and significance of not pushing our own development without bringing along with us the less-favored areas of the world. It will do us little good to realize the promise of the nuclear age within our own borders unless, at the same time, the well-being of other nations is improving too.

Millikan and Rostow, in their book "*A Proposal: Key to an Effective Foreign Policy,*" discuss some of the aspects of what this means. It is the burden of their argument that there are three stages in the growth of societies:

> First, there is a period during which the preconditions for economic progress are established. . . . Following this precondition stage there comes a second stage during which the country makes the complex transition to a position where sustained economic growth becomes possible. . . . Third comes the long period of regular if fluctuating progress. . . . The American purpose would be to assist

countries in establishing the preconditions for growth (mainly through technical assistance) and to assist them with capital, notably in the decisive period of transition. The objective is not to redistribute income as between the industrialized and underdeveloped nations: the objective is to help the underdeveloped nations into the stage of self-sustained growth.[3]

Among the tasks is the use of our knowledge to shorten the process by which in the past nations have moved from the first to the third stage. For this to be accomplished any national policy we formulate to this end requires the support of public opinion. At the same time public opinion quite naturally is concerned with the problems and possible advances it sees at home. Unless political leadership can give effective expression to the link between our own projects for greater well-being and the needs of other peoples, the natural course is for us to give most of our attention to the affairs of our own communities and our own society.

There is a strong body of informed opinion that holds that this, in the long run, would isolate us in the world and drive less favored peoples into the Russian orbit. It is not the task of political communication to judge or argue or advocate one decision as against another. But it is important to notice that political communication will be profoundly affected by whatever decision is made.

We can hardly expect to influence, and draw into our circle, nations who see us developing great national prosperity with relatively little concern for the problems they face in attempting to pass from the "precondition" stage Millikan and Rostow talk about to the stage "where sustained economic growth becomes possible."

Political communication by the leader of the free world will be difficult, if not impossible, unless, in a pragmatic,

[3] Max F. Millikan and W. W. Rostow. (First ed.; New York: Harper & Brothers, 1957.) Pp. 44, 46, 47, 54.

realistic way we assume some measure of responsibility for being our brother's keeper. In this context it is not enough to cite the Marshall Plan, the work of the International Cooperation Administration, our programs of military aid. It is not reactive, short-range devices that are needed, but a long-range, stable, planned, and continuing concern with the problems of international development along lines that encourage and support the freedom of men and nations that will carry conviction to the observers of our policies.

That there is, in fact, this realization by leaders of opinion and makers of policy is supported by many items of evidence. Our diplomatic, economic, and military instruments of government, on the record, have demonstrated a growing interest and skill in facing it. But it is being faced in our own way, in terms of our own political institutions and practices, and these are not always comprehended or understood outside our borders. An instrument for the explanation and interpretation of our own methods and actions, and for the interpretation of our own learning processes as we face, for the first time in our history, the responsibilities of international leadership, is called for. This means that we must concern ourselves with ideas, and the meaning of ideas, quite as much as with the means for expressing those ideas in concrete form.

This recognition has been accorded by the Soviet apparatus of government. Ideas and aims and objectives and national policies are completely integrated in the political maneuverings of the Russian leaders. And the effectiveness of their system for influencing the minds of men is attested by the position and role they occupy. It is the conviction and contention of the United States—its people and its government—that the premises on which the U.S.S.R. is proceeding are, in important respects, false and corrupt. They are sufficiently seductive, however, to require us, if we wish to discredit them, to take them seriously and combat them by the mobilization of our entire resources. The one instrument we have not yet considered seriously enough in this conflict is the instrument of political communication.

Let us look at its status in our arsenal at the present time.

In the first place, we are still thinking of it as psychological warfare. It is true that in our practice we have gradually been moving toward a less restrictive employment of it —the highest levels of government now recognize that in any decision on national policy there are psychological factors that must be taken into account, but we have still to reach a position on which there is common agreement about the nature of the instrument.

There are still two major positions that are unreconciled and unresolved, and both have powerful adherents and advocates. On the one hand, there are those who regard the instrument as a means to help in the attainment of national objectives by the use of both trickery in propaganda and, when it serves our purpose, undistorted or unslanted facts. Thus, for example, we have Radio Free Europe reportedly solemnly taunting the Russians in radio interviews with fictitious characters.[4] On the other hand, we have those who take the stand that our use of this instrument must rest on the truth as truth has been defined in the first of the basic premises discerned in the second chapter of this book.[5]

Self-evidently we cannot afford to have two such conflicting assumptions. We must decide on one or the other and stick to it. Furthermore, the decision has to be one that will receive majority support, and it must be one that is made in terms of analysis and rigorous examination. It will not do to hand it down by fiat.

In making up our minds about this instrument, it is well to remember that if, by our practice, and by public discussion, we show that we finally have elected to wage psycho-

[4] Ch. II, footnote 51.

[5] See p. 26.

logical warfare, we cannot avoid having our friends, and neutrals, harbor the suspicion that our activities in this field cannot be taken at their face value. There must always be the feeling among them that perhaps what we are saying has an ulterior purpose, and it will be this purpose they will seek, and if, in truth, such a purpose is discovered, our credibility will suffer.

In fact, looking at this instrument today, it is difficult not to conclude that our organization for its use has yet to determine, and agree on, what it is we want it to do and what the ground rules shall be. Looking back over the recent years, it seems quite clear what the U.S.S.R. expects from it, and their ground rules also can be discerned from their practice. Our own recent history shows that we have suffered from not being clear in our minds as to what the activity is all about. Our enemies could make a good case for arguing that we have been unprincipled in its employment. Our friends could discern a well-intentioned beginning, but they would surely have to record lapses in judgment and errors of omission and commission, resulting from conflicting premises.

If we are to give the instrument an opportunity to show its full capabilities, we ought surely to take a stand that, on the basis of the record as we employ it, will show quite clearly the principles that govern its use.

In the second place, the instrument is obviously unco-ordinated. There are patent difficulties in the way of accomplishing this in a free society, if for no other reason than that such a society only rarely speaks in a united voice. But the confusion could well be lessened by central direction and, again, by recognition of what the role of the instrument is. Clearly there cannot be anything but confusion if its role is thought of as that of holding up to the international gaze the kaleidoscopic American scene, in the hope that those who view it will be able to determine for themselves those trends in it that are significant and those that are evanescent. But if that role is taken to be the explanation and interpretation of government policy, there is opportunity for less diffusion in focus. Not that we can as yet say that our policies do not fluctuate, but their sum, looked at over recent years, moves unmistakeably in a definite direction.

It will not be possible to co-ordinate this instrument until agreement has been reached on its role. This role is not departmental. There is no place for a department of political communication in the sense that there is a place for a Department of Defense, a Department of State, or an Office of Defense Mobilization. The task of political communication is to take the activities of such departments and interpret and explain them so that they give as clear and coherent a picture as possible of what our national aims and objectives are. The role of the instrument is, first of all, national. It is therefore above departments in the sense that no single department can be said to speak with a national voice. It is the sum of the activities and decisions, the compromises and debates of the great branches of government that ought to be interpreted and mediated to peoples beyond our borders. This sum cannot be reflected by any of its parts. The primary concern of the Department of Defense is war and our survival if war comes. The primary concern of the Department of State is our foreign policy and the diplomatic machinery that is necessary to give it substance. Our national objectives are served by both departments, among others. But neither, by itself, can determine national policy or elect the compromises that national policy may find desirable. It is the constant adjustments reflected after the departments and machinery of government have considered their positions that determine our national policy, and it is this policy that must be interpreted and made clear to the world. No department within government can make this synthesis. It requires an organization speaking for departments, but not itself departmental.

At the same time the established departments of government clearly possess what may be called subroles in the field of political communication, which they can properly fulfill in departmental terms, but meaningfully only if they are subject to central, co-ordinating guidance.

Thus, for example, in war, combat psychological warfare is the proper concern of the military establishment, and it ought to be carried out by its own officers and men, especially trained for the task of addressing an enemy. Similarly, at times other than war, the necessary information services of a government are the concern of an agency equipped to provide foreign people with proper knowledge about our society, its culture, its political practices.

We have not, as yet, reached an agreement on the nature of the national role of the instrument nor the nature of the relationship and responsibilities and tasks of the agencies with subroles. Thus we have still the unresolved problem of how we should undertake political communication in a military theater of operations. The Department of State views this kind of operation as strategic in nature with effects going beyond the end of active hostilities. Therefore it considers that not only policy but certain actual operating responsibilities should be under its jurisdiction. The military establishment agrees on the strategic aspect of the task, but is firmly opposed to any but military control of such operations.

In fact, the national nature of the instrument ought to be the decisive factor in any decision. No theater commander can make national decisions. These are transmitted to him through the Joint Chiefs of Staff. With isolated and magnificent exceptions, which serve only to support the general observation, no theater commander would think of trying to dictate national policy in a war or draw up the political terms on which the nation would accept surrender. Yet when it comes to trafficking with ideas and attempting to influence men's minds, theater commanders in the past have had no hesitation in dictating what should be said and what should be done. In practice, repeatedly, the control of operations has led to predominance in the making of policy.

Because of its nature, which requires political communication to be in terms of national aims and objectives, the instrument ought not to be under the control of a military commander in time of war. The machinery that will permit the instrument to function in terms of its true role, while yet being answerable to a commander for not threatening the security of his military operations, is neither difficult nor particularly complicated to devise. It can take several forms. The prime ingredient is good will. But in the absence of general recognition of the national role of political communication, differences persist throughout the hierarchy of government and prevent the instrument from fully effective performance.

Furthermore, in looking at its present status in government, it is undeniable that there are problems about parts of the machinery for political communication, and that there are good grounds for contending that sufficient knowledge exists on which to base executive decisions. But these decisions are not being made at the level required, again because the national role of the instrument is not generally accepted.

Intelligence is one of these problems. Personnel is another. Research is a third. All three of these problems are vital to an effective operation. We have now a fund of true, transmissible knowledge about them sufficient to permit some executive decisions. The decisions will have to be executive in nature, for the recent history of our activities in this field shows clearly that the various departments that have been chiefly concerned are unable to put themselves on the team. The positions taken in the past have acquired the patina of precedent and, although in age these positions are young, they have become encrusted and surcharged with emotional overtones.

For example, the personnel presently engaged in forms of political communication undoubtedly will need to be augmented if hostilities on any major scale develop. It is unlikely that there are enough men to fill the requirements. On what basis and in what proportions shall present personnel be divided among competing needs, and on what criteria shall augmenting reserves be recruited?

In World War II the competition among departments and agencies of government for men with qualifications in this field was open and unregulated. We have not sufficient skills and talents to go around under such circumstances. We do have now a body of experience and knowledge that could be applied to a solution, which would be preferable to waiting for the need to arise and making *ad hoc* compromises.

Research is one of the vital and principal tools for effective political communication. It is the disciplines of the social sciences that seem to be most urgently of significance. Yet, relatively speaking, very little money is being given for research in this area. Nor, on the basis of past experience, is a large sum called for. An increase on the order of two- or three-tenths of 1 per cent in the national budget devoted to research and development, guaranteed for five years, would in all probability show whether or not these disciplines were capable of making significant contributions to our knowledge of how most effectively to interpret and explain our national aims and objectives to international audiences.

The problem of intelligence calls for special handling, and the requirements here will be discussed in the next section of this chapter, when certain specific actions that appear indicated are suggested.

Action on these and similar problems calls first for co-ordination. As our political institutions have developed, it seems that less and less have the great branches of government become able to speak in national terms on the great issues of policy that arise from time to time. Increasingly it is to the White House that the country looks for the expression of a national position, for the resolution, in national terms, of divergent courses of action. The quality of leadership required of a President under such circumstances has been magnified by the complexity of modern problems, by the necessity to choose and elect from among a variety of possible decisions the ones most advantageous to our aims and objectives, by the knowledge that the speed and range of technological progress denies us the opportunity to test and try all available proposals. Decisions must be selective in nature, for the sheer cost of developing each and every proposal to the point where it can safely be determined whether or not to pursue it further is prohibitive.

Decisions of this nature, fraught with the possibility of national disaster if in important respects they are faulty, place on the White House a burden beyond the capability of any one man. They have therefore to be delegated. It is combined consideration and combined judgment that is called for, but it is the President who, after such consideration, must elect the decision and shoulder the responsibility.

The nature of political communication, since it is the explanation and interpretation of our national policies and our national aspirations, also, in all probability, requires direction at the level of the White House. Political communication is a full-time task in its own right and therefore beyond the power of the President himself. Nevertheless, the record of our history, our experiences with what we have called psychological warfare, and the fact that unresolved differences of opinion have prevented us from formulating an acceptable, workable organization and plan that is supported by all the branches of government seem at this juncture to call for executive decisions and executive action.

Only if the President calls for such action, and only if the President endorses the decisions, does it look as though we shall be able to put this instrument to work as effectively as an analysis of its potentialities indicates is possible. This seems to mean that in this field as in others, the President must, in effect, delegate to a selected member of his staff some of the powers and authority he wields by virtue of his position. There is a clear requirement, in other words, for a co-ordinator of our political communication who is above the conflicting viewpoints of departments and agencies in this field.

The caliber of man, and the quality and character of his attainments, are obviously rare. C. D. Jackson, who for a time was the President's Special Assistant for psychological matters, in conversation with this writer discussed some of them as a result of his direct experience.[6] In the first place, the task calls for the authority necessary to resolve issues in this field on which departments are unable or unwilling to come to operating decisions. This resolution obviously can only be made by the President. The powers, prestige, and position of the departments concerned are such that it is unlikely that they would accept anything less. It seems to be currently agreed that if the President acted on some of the issues outstanding—notably, for example, the issue of control in a theater of military operations—the decision could be made to stick.

The decisions, themselves, however require analysis and thought and familiarity with the history and reasons for

[6] Private Memorandum of conversation, April, 1954.

departmental positions that have grown up. They are not simple. They call for good will and the ability to engender good will. In the last analysis, they call for firm decision resting on insight and knowledge. No President, himself, can bring these qualities to the task along with all his other duties. The task must be delegated, with the President affirming the solution over his own signature.

In other words, there must first be the attempt to bring about the resolution of differences by the departments and agencies concerned themselves. Behind this attempt must be the certain knowledge that if this proves impossible by reason of intransigeance, the recourse to the President will bring instant, positive action.

Among other qualities, such a co-ordinator will need status and prestige, and these can be conferred only by the President, for a co-ordinator, in view of the history of this activity, will need at times to confer directly with the heads of powerful departments. He will have to confer on terms of equality. In part, this equality must be the result of his own abilities and caliber. It must be clear that his qualities of mind, character, and achievement entitle him to the position. At the same time, this ability to move horizontally along government levels must be matched by both the desire and the ability to move vertically.

This vertical movement will take him down from the highest levels of government to the working levels, where he will meet the men who are engaged in the day-to-day operations of an organization for political communication. It is at these working levels that the way policy and operating decisions at high levels influence practice can be observed, and it is at these working levels that frictions start and unworkable or faulty decisions show up. Pre-eminently in this field there is the requirement for walking with kings, yet keeping the common touch. But always present must be the knowledge on the part of all whose work brings them into contact with the machinery of political communication that the co-ordinator has the right of immediate access to the President, and that this access, on issues of moment, brings swift action.

It will in all probability have to be through the authority and good offices of such a co-ordinator that the problems of intelligence, personnel, and research, among others, are satisfactorily dealt with. Solution in these areas is not just a matter of seeing that directives are issued. The nature of political communication is such that frequently the issues require separate consideration each time they arise and a decision on the merits of specific instances. One decision, once taken, is not, as often is the case in other activities, sufficient to guide future developments. Access to certain intelligence is often indicated in one instance and counter-indicated in another. Decisions in the name of security are necessary and desirable under some considerations, while, more rarely, security ought to be waived in others.

Analysis and examination of the problems of political communication, together with the potential importance of the instrument as a means for attaining our aims, seem to show quite clearly that only under some such co-ordinator can the existing agencies of government, responsible for phases of the national operation, be welded into a unified whole, serving the national interest of the United States and making clear to the world what it is we seek and what relevant action from them is both consonant with their own interests and agreeable to ours.

Such a position will carry with it, inevitably, tremendous power. Herein is one reason that supports the creation by the Congress of a joint committee authorized to become privy to all that we undertake by means of political communication in both our white and our black operations. Many of the activities in this field are secret and properly so. Because of this, many differences of opinion may not be aired for the beneficial astringencies of public opinion to perform their annealing function. We may confidently predict that differences of opinion will persist, whatever machinery we devise to permit this instrument to function more effectively. One of the important functions of a Congressional committee would surely be that of a safety valve in that it would allow differences to be voiced and considered. It would also—and this, considering our traditions, is important—act as a powerful brake and check on the opportunities for a co-ordinator to exercise too great a personal power and cherish unwanted and dangerous ambitions.

If the attempt that has been made to describe political communication, and to describe the position in which it finds itself, is correct in the essential particulars, it is proper to consider the several courses of action that appear indicated if it is to take its place alongside the diplomatic, economic, and military agencies of government as an effective instrument of state.

First, there seems to be a clear need for a co-ordinator. But there is no need for the creation of another department of government to undertake political communication. The task, in essence, is one of winning acceptance among the existing departments of government for the concept that the sum of the actions for which they are responsible makes up the voice that is heard by the peoples and nations of the world. Political communication is concerned with the meanings that can be given to this "voice," and the prime task of a co-ordinator is to see to it that the appropriate agency for interpretation undertakes the task and, in discharging it, is not in unnecessary conflict with other aspects of political communication going forward by brother agencies.

This co-ordinator, then, whatever may be his title, will need to be aware of the operations and plans of the United States Information Agency, the military establishment, and the Central Intelligence Agency. There will be strong and powerful opposition to this by many able and experienced officials in government who will not relish the authority of such a co-ordinator. The departments of government are jealous of their prerogatives and powers, and properly so. In the discharge of their functions they tend to recognize only one overriding executive control and that is the White House in the person of the President. The objection will take the form of arguing that a co-ordinator with such authority will, in effect, be above such important cabinet offices as the Secretary of State and the Secretary of Defense. This, the argument will run, limits the right of access of such officers to the President on matters involving political communication. This will be intolerable. But also, the argument continues, a co-ordinator who is not above such officers, but only ranks with them, will not be able to co-ordinate. The conclusion of this argument is that the appointment of such a co-ordinator is politically infeasible.

The argument is cogent and has much force. An answer to it depends on acceptance of the concept of political communication. If the function of the instrument is national in scope, and not departmental, it would seem that the only lodgment from which it can be directed is the White House. It is impossible for the President, himself, to give it the necessary daily attention. In such circumstances the only recourse is a delegation of his interest and concern. Clearly this delegation cannot imply that the responsibilities and powers of heads of important departments are being abrogated. In fact, however, under the burden of departmental affairs, such heads, themselves, are unable to give much time or thought to the problems of this instrument. Nor are they equipped, by experience, training, or practice, to comprehend the peculiar nature of its problems, needs, and capabilities.

Much depends upon the character and quality of the individual selected to function as the President's right hand.

If he is able to win acceptance of his position by the manner in which he undertakes his task in his meetings with the heads of departments, he will, if he is versed in the activities of his field, be able to win departmental support by the logic of the courses he suggests for implementation. He will not be able to exercise his role by fiat. Success and influence in the direction of political communication will follow from demonstration. He will need to be able to present his case at times against what Forrestal has called "an earnest and intelligent opposition."[7] He will need to proceed slowly, for the task is not one simply of co-ordination, it is also one of education, of showing by example the nature of the relation between the actions of government and their explanation and interpretation to alien peoples.

In effect, the authority and influence and weight he is able to exercise depend largely on his own ability to persuade and inform his colleagues and on the extent to which he is careful never to force decisions by virtue of his position unless, in default of decision, our political communication is estopped from interpreting our national aims and objectives.

There does not seem to be any need for such a co-ordinator to preside over a large staff. He will need and should have a few competent aides from the departments and agencies presently concerned with aspects of our political communication—officers from the military establishment and representatives from USIA and CIA. It is by such a small group that he will be kept informed of the individual courses of action these agencies are following. And it is through such a small group that national policy decisions can be given appropriate weight and emphasis in our messages.

The difficulty of the task of such a co-ordinator will be

[7] Walter Millis (ed.), *The Forrestal Diaries* (New York: The Viking Press, Inc., October, 1951), p. 92.

increased by the fact that he will not be able, by virtue of authority alone, to direct or override the officers in charge of our present activities in this field. The director of USIA will still be the director. The activities of CIA will still be the responsibility of that agency. But it will then be known where they are in conflict, and any duplication will come to light. And the machinery for interagency support of the national objectives will be provided, for, once it is clear that the co-ordinator is backed by the President, conflicting actions and conflicting positions among the present agencies will tend to be resolved, since, if this is not done, it will become clear that action by the President will follow.

The task of co-ordination in the field of political communication, in other words, seems to call for the skills of negotiation and creative imagination, and while the office probably cannot confer direct authority for enforcement, it must be clear that Presidential action results if there is a lack of co-operation or intransigeance among the members of the national team.

It follows that such a co-ordinator must be a member of the policy-making group in an administration. If he is to try to give force and meaning to the actions that are determined as policy, he must, necessarily, understand the reasons that impel them. The interpretation of a nation's policy is not just a matter of knowing what the decisions are. How they were reached and how differing points of view were reconciled or compromised is at least equally important. It is Lovett's opinion "that foreign policy, like a war plan, could not be stated in a single speech . . . but was a planned, continuing flow of actions and statements directed toward a central objective."[8] Finally, it is at the time decisions are made that the psychological implications should be considered. These should not emerge for what use can

[8] *Ibid.*, p. 443.

be made of them afterward. The actions of government can always be given meanings, which is only another way of saying that the psychological aspect is omnipresent. It may, at times, be necessary to disregard it or to run the risk of unfavorable psychological reactions. But the time for all of this is when the policy itself is being determined. And the co-ordinator of our political communication is the logical official to present such facts as may be known in the area of his cognizance bearing on the decision.

So far, the history of the activity shows this has not been done. And for lack of it, repeatedly our information activities have been inept, or made to look ridiculous, to the discomfort and ensuing strain on morale of those with the operating responsibilities.

One of the first tasks, if not the first, facing such a co-ordinator, if he is appointed, is the resolution of the long-drawn-out differences of opinion between the Department of State and the Department of Defense over the way strategic political communication should be directed in the event of war. The roots of this problem go back all the way to World War I, when, in Europe, the Creel Committee assumed responsibility for propaganda to civilians, and the Army assumed responsibility for addressing combat troops.

There can be no question that combat psychological warfare is the responsibility solely of the military establishment. If this were all there were to the problem, it would have been settled long ago. It involves much more than this, however, for the nub of the predicament we are in today centers on varying interpretations of what is meant by strategic activities in this field. The key to any solution seems to lie in the basic premises governing political communication undertaken by a free society. If those premises have been correctly inferred and set down, then political communication at all times needs to be under the direction of political as distinct from military authority. And in time of war, to an enemy, strategic psychological warfare, which is a branch of our political communication, is answerable in the same manner to those premises.

It also seems true, however, that political authority alone is insufficient. One of the clear and patent facts of the age we are entering is the necessity for military partnership in the determination of national policy. This partnership is required as fully in the determination of our strategic political communication as it is in the determination of the other aspects of policy. The significant point is that the nature of this partnership by the military cannot be the controlling factor. True partnership is required, but not controlling partnership.

It is the strength and essentiality of this position, when political communication is undertaken by a free society, that must be kept clearly in mind when approaching the problem of reconciling the differences in this field between two such powerful departments as State and Defense, both of which by now have encamped on positions that make it difficult for either to retreat with good grace.

It is probable that, if we are to make effective use of this instrument in war, a decision will have to be made by the President as Commander-in-Chief of the Armed Forces. However, such a decision without preparation would in all probability not accomplish its objective. As President Eisenhower said at the time of the unification hearings before the Senate, good will is the essential component.[9] The first

[9] President Eisenhower, then General of the Army, at that time observed: "In the vast human effort, goodwill is far more conducive to efficient operation than is any mere organizational detail. To secure the utmost in eventual effect the original plan should go no further than to establish basic principles and to prescribe the plan so far as applicable to the area of general agreement already obtained." Hearings before the Committee on Armed Services, National Defense Establishment, "Unification of the Armed Services," United States Senate, Eightieth Congress, First Session, S. 758, p. 90.

See also James V. Forrestal's strikingly similar observation: "Goodwill

step would seem to be a determined attempt to achieve that good will without which no acts of Congress or executive orders can bring forth harmonious co-operation.

A further problem a co-ordinator faces is that of intelligence for political communication. We do not yet know, in enough detail, nor with sufficient authority, what kinds of intelligence the instrument needs, nor the best method for providing it. There are good grounds, however, for taking the position that we have the knowledge available, although as yet in uncollected but transmissible form, if a concerted effort were made to find it.

It is probable that this attempt—it belongs in the domain of research—ought not to be undertaken by any of the existing organs of government. Like the problem between the Department of State and the Department of Defense, the various intelligence agencies of government have taken up positions that are backed by emotional feelings as well as historical precedent. And if the usual reactions of human nature apply, it will be difficult for them to divorce themselves from subjective, and even subconscious, stumbling blocks in trying to discover the best way to serve the voluminous and catholic requirements of political communication. If we are to find out both *what* intelligence is needed, and *how* to provide it, the task probably ought to be undertaken by an independent research agency at the direction and request of government at a high level. Furthermore such an agency must, itself, have some awareness of the general problem of intelligence; it must, in other words, approach its task with sophistication as well as authority to get the information it will need. This limits the research

can make any organization work. Conversely, the best organization chart in the world is unsound if the men who have to make it work do not believe in it. I am not naive enough to suppose that legislation, no matter how excellent, can insure the cooperation and goodwill that I consider so essential. These attributes come from the heart rather than from the mind and therefore are beyond the scope of legislation." *Ibid.*, p. 23.

facilities available, but there are several agencies that qualify and whose record in working in government with both integrity and discretion is attested.

The problem of intelligence is of vital importance to the success of any program of political communication, and it is one that calls for quick and determined action. This writer takes the position that an answer could be found if only sufficient interest and initiative were forthcoming at a level that would ensure the necessary administrative support. The problem is one that would benefit from the sympathetic and powerful direction a co-ordinator could bring to bear on it.

Two further problems, which appear to need the guiding hand of a co-ordinator if an appropriate answer is to be found, remain for discussion. The first of these is concerned with the establishment of a college that will teach the principles of political communication. This is different from, and should not be confused with, a training college. It should be possible, if mature consideration indicates the desirability, for such a college to include certain training courses for those actively engaged in the operational aspects of political communication. But, in the writer's opinion, the reasons are more compelling for considering that such an institution should place primary emphasis on the role of teaching. It is not so much the practitioners of what we have called psychological warfare who need training, as it seems to be the officers of government in all its many branches who require teaching in the relation of this instrument to the already established instruments of government, which are accepted and with which they are familiar.

There is much to be learned here. We are barely on the threshold of knowledge and understanding about the forces at work in the world today, about the pressing requirement for facing the tremendous problem of what conceptions, rooted in our history, need now to be rethought or discarded in terms of more exact knowledge gained as the price of admission to a nuclear age.

Such concepts as the nation-state, sovereignty, and the emotional forces conjured into play by nationalist tendencies, on the level both of the great powers and lesser powers just now emerging from "backward" or colonial status to a position of independence and "self-determination," require searching analysis and thought. They are at the heart of political communication.

It is not only our pragmatic actions and policy decisions that require mediation to these powers and to the world. There would seem to be a patent requirement, at the same time, to communicate and share with them the reasons and motives by which we are impelled. It is a basic assumption in this book that these reasons and motives are not solely selfish and purely national. If this assumption is false, there is no point in considering political communication as an instrument of state. Instead, what is called for is the weapon of psychological warfare. But there seem to be good grounds for taking the position that the American people are in fact tinged with idealism, and that the concept of democracy, as our history shows it to have been practiced, permits the conclusion that we are not altogether comfortable about the circumlocution "enlightened self-interest" and that, in fact, we mean by this that we are really concerned with the welfare of peoples other than, as well as, our own and would indeed be willing to help them on grounds of humanitarian concern provided this same concern were shown by them in their own relationships.

We are, as Karen Horney has said, ridden by two devils, one of whom exhorts us to brotherly love while the other bids us to look out for number one. The point is that we are ridden by *two* such devils and not only by the second.

It is this uncomfortable knowledge that is responsible for certain aspects of our appearance in the eyes of the world. It is the fact that as a nation we are still willing to go to church on Sunday, whatever lapses may be called for during the week, that puts us in the position of showing that we are manifestly concerned about others on the one hand, while at the same time employing the goad of self-interest to prod them in a direction that we believe is in their interests and at the same time serves our own desires.

This conflict has been present ever since man began reflecting on his nature. The point that appears significant now is that our rapidly increasing knowledge is facing us with the task of attempting to find the nature of the relationship between the spirit of man and his material gains in a modern world and giving this expression in terms that make possible some sort of unity among seeming incompatibilities. There is no such requirement from the ideology of communism as it is being practiced. It is implicit and inescapable so long as men seek knowledge in terms of a society that is their servant and not their master.

The fact that demonstrably we have now reached a stage in the development of societies and knowledge that links our political, economic, and military activities in a way that they have never been linked before, making all decisions interdependent, appears to point up strongly the need for ideas and concepts that will win acceptance as guiding principles for international relations.

The relatively simple political and religious concepts of the nineteenth century, which so many seem still to accept, no longer apply to our societies as our universe shrinks, both timewise and spacewise. But the concepts that must replace them have not yet been expressed in terms capable of being grasped by large numbers of mankind.

There is ready recognition of this by communism, and it is the seeming inability of the West at this crisis in history to give reasons for "the hope that is within it" that allows Communist leaders to increase their influence. For it is the negativistic ideas of communism that are carrying appeal: the idea that Western materialism is selfish; the idea that the patent inequities of Western civilization are deliberately designed to prevent the growth of prosperity among the people of all nations; the idea, which takes subtle advantage of the concept of privatization, that the Western leaders are manipulating the forces that make for economic progress and stability in order to preserve an elite position.

Essentially these ideas are negative in aspect, but they are clearly powerful levers. In opposition to them is the basic concept of the West: Ye shall know the truth and the truth shall make you free. What this truth is we do not seem as yet to know, or at any rate to be able to express, in terms that carry wide appeal. Yet is the West instinct with it. It is an idea that history shows we have believed and for which we still seek knowledge. Our difficulty is to preserve the form of society that will permit us to seek still greater knowledge while being able to hold and win the allegiance of men to a concept that as yet, in twentieth-century terms, we are unable to express.

This is, in truth, a conflict of ideas and on a scale greater than we have faced in our history so far. And this implies that the conflict must be fought at the level of ideas as well as at the levels of diplomacy, economic action, and military preparedness.

The role of a college that will search for and teach the nature of the quadripartite relationships of nations and societies today—the interdependent harmonies necessary among our diplomatic, economic, and military courses of action and the role of political communication in giving them positive interpretation—needs wise and careful consideration by leaders of our society in all these fields and requires the attention of a co-ordinator able to bring these leaders together to give the problem thoughtful reflection.

The other problem centers on the requirement for co-ordinated research in the field of political communication. So far, research in this field has been undertaken by certain agencies or groups for specific clients—for example, the Army, or the Air Force, or the United States Information Agency. Such research programs have rarely, if ever, been discussed among, and by, all the agencies concerned with political communication, to determine whether or not the research problem was of wider application than to the needs of a single client only.

In many instances this wider application is obvious to any student or practitioner of experience. Clearly it would be in the national interest, in such instances, for all the agencies to whom the problem and its solution were significant to initiate action in concord. Not only does this not take place as "standard operating procedure," but, what is worse, the results of research have a hard time getting general distribution so that those who are operating in this field know what gains are being made in knowledge and in what direction those gains are moving.

For example, out of a number of "messages" that might be addressed to an "audience," which one has the greatest likelihood of being effective? An answer to a problem of this kind obviously is just as valuable to the United States Information Agency as to the Army. The necessary techniques and knowledge are of importance to both military and civilian specialists in this field. Under existing circumstances, either military or civilian agencies concerned with aspects of political communication might initiate research activity. It would be highly unusual, however, for them to discuss the problem with each other and with the research agency tackling it, before work got under way, in order to prevent duplication of work or to determine whether aspects of the problem of concern to one or the other might not be included, so as to make the work of widest applicability and value.

There is another aspect of the problem, which is the direct outcome of bureaucratic government. It is at times clearly necessary, in attempting to increase our knowledge, to consider related fields not specifically under the jurisdiction of the department of government requesting the research. If the activities of other departments should be discussed, analyzed, or examined in order to determine more accurately the nature of the solution necessary, it is likely that the resulting research document would have a hard time getting distribution in government circles, particularly if that part of it relating to departmental activities outside and beyond the cognizance of the originating department is critical in nature.

The conventional position is for departments to stick to their own knitting and not concern themselves with the work and activities of their counterparts. This is perfectly natural and understandable when the problems are truly limited in scope and peculiar only to specific departments. But the problems of political communication are not of this nature. They are, generally speaking, common to all agencies having any concern with operations in this field. There are few problems of research that are of military significance and interest only. Not many questions that USIA might tackle have no interest to the Army. Thus, effective research in the field of political communication ought to be of common concern and commonly discussed. And any answers gained by one ought to be available to the other. Provided a research document does not go "hunting" in the territory of a department not responsible for its "license," distribution may be expected. But if, so to speak, the covey has gone across the fence into another field, and been followed, this is trespass, whether or not the resulting gain in knowledge is likely to be of value.

For two reasons, what is required is co-ordination of research: first, because the problems cannot be departmental;

and second, to avoid duplication and save money. Like the activity, the problems are wider than departments and are more often of general, rather than departmental, interest and significance. Co-ordination of research ought not to mean that a central office determines what research should be undertaken and what research agency would be responsible. It is of the essence in research that results be available for critical and rigorous examination by independent specialists. This is a vital safeguard against error. Thus, operating agencies, and the research units serving them, ought to be free to determine the research they will undertake. What is required is for all operating agencies to be aware of projects under consideration, so that aspects of interest to them may be included or given specific attention and the widest possible scope covered.

Beyond this is the strong indication that research in the field of political communication is at its best when done in the laboratory. One of the lessons of our experience with psychological warfare in the Korean war seems to be, quite clearly, that the opportunity to undertake research on problems arising in the field yielded important results. The best laboratory at present is provided by the on-going work of the United States Information Agency throughout the world. But this laboratory is not available to the research agencies serving other departments of government that are interested in aspects of political communication. Furthermore, there are results from work done in the Korean war that need validation by repeated experimentation. This experimentation can be done only in the laboratory of an on-going operation. Validation of some of the work done in Korea would be of value to both the military establishment and USIA. But until there is co-ordination and effective, sympathetic liaison, it is unlikely that it will be done.

Among the first of the important tasks of the small staff supporting a co-ordinator ought to be the co-ordination of research, meaning by this, seeing to it that all research projects are known and discussed by all operating agencies before being initiated. Second should be the establishment of machinery for making available to practitioners the results of research in terms they could understand and apply. And the third task should be the determination of priorities of research in terms of the national nature of the instrument, as distinct from the purely departmental.

To accomplish this, representatives of the research and operating agencies should meet under the chairmanship of the office of the co-ordinator. The problems, priorities, and the action to be taken ought not to be determined *for* these agencies, but *by* them, with the co-ordinator's office acting only as the catalyst bringing them together and taking action itself only if the representatives present were unable or unwilling to co-operate.

It is time to consider "the conclusion of the whole matter," for, in truth, "of making many books there is no end; and much study is a weariness of the flesh."

If we are honest, it seems to this writer we must admit that the record of what has been called psychological warfare, and of what this book calls political communication, is not particularly impressive. The evidence does not indicate that this is because the instrument is ineffective. Rather, it indicates failure to understand and apply the principles by which the instrument is governed in a free society.

The important question is not, therefore, whether or not we need the instrument, but, rather, what the requirements are if it is to serve us in this time of peril. The force and relevance of ideas are shown by the history of man's development. No one can deny that the ideas of the West have been powerful in the past. Do these ideas still possess power to win and hold the allegiance of men? Is the Western heritage crumbling under the impact of increasing technological knowledge in virtually all the fields of man's endeavor?

Assuredly increasing knowledge has undermined ancient concepts and traditions that we have been accustomed to accept without thought. Has it struck down the root concepts on which a free society is founded? If not, then it is the expression of these concepts and ideas in terms that can be related to the hopes and aspirations of men that seems to be important.

The instrument for this is political communication. Military strength, economic power, diplomatic skill are means to ends. In themselves they are inherently neither good nor bad. Ends, in the long run, are embodied in ideas. It is what nations will do with power and how they exercise it that attracts or repels.

The ideas on which our society rests are common knowledge: the equality of men and their right to life and liberty. It is not enough merely to defend these ideas against attack, which is what we have been attempting to do. Increasing knowledge calls for constant reassessment of the ideas and, if they stand up, the re-expression of them in terms of such knowledge. Then there is required militant explanation and interpretation of them as they can be shown to govern the decisions involved in our national policies.

It is the spirit of men with which political communication is concerned and the way this spirit is expressed by the sum of our diplomatic, economic, and military decisions and actions. This sum, self-evidently, is not perfect. There are conflicting drives and ambitions and corrupting influences. But it has not yet been possible to show that the heritage of the West looks to man's enslavement. In other words, the key to our traditions is the presence of hope and faith that men, given knowledge and accurate interpretation of what that knowledge means, will combine to achieve its promise, not by force, nor by fiat, but through co-operation.

That we may fail is undeniable. That we may surrender in the face of catastrophe is possible. The third alternative is that we face the challenge soberly and in unity, aware that the issue of the struggle, unless we slip into all-out war, is, in all probability, far in the future.

If we elect this third alternative, we have no option but to give constant expression to our hope and to show how and why our actions support it. That we shall fail in this communication at times is inevitable. But that this failure will destroy us, and therefore destroy freedom, is unlikely on the basis of past evidence. It is not the failure to attain their highest and deepest aspirations that has led to civilizations' falling, so much as it is contentment with an existing stage and failure to improve it.

The first essential is national leadership, a leadership that can light up the wall of premises behind a free people and give those premises meaning and significance in terms readily comprehended in the age we face. The second essential is the explanation and interpretation of the actions governed by those premises. For this we need active, militant political communication. The weapon is on the wall when it should be in our hands.

Appendix

To the Soldier Who Marches West

You are still alive. It is wonderful. Everything that lives is wonderful, even the green grass and the birds.

The dead and the rocks and the soil and the dung—they are nothing, for they have no life.

We who have life have everything; we possess fabulous wealth.

The rocks, the dead and the soil have nothing, are nothing.

Where will your road lead you, soldier? Are you going West? Are you going to Paris?

Do you know what is in the West, soldier? I shall tell you, soldier; listen.

In front of you are the English, you know that. The French and the Americans are behind them. You also know how they fire into the lines of your comrades. Perhaps they will retreat and new regiments will march ahead. Then they will fire again. Then the Allies retreat again.

But the firing never ceases. . . .

There is something else in the West. I shall tell you what this "something" is. Nobody can tell you exactly where it is, but it surely is in the West.

Your grave lies in the West.

If you march West, you can't help finding it. Possibly it is far ahead of you, behind the mountains. But possibly it is very near you, perhaps you can see it. Today or tomorrow —nobody knows. But surely, the grave lies there, as surely as does the sunset.

Do you march West, soldier? Then, we say good-by to you. All of us who live say good-by.

There are only two things on earth, the living and the dead. The difference between these two things is greater than that between friend and foe, greater than that between man and animal. It is the greatest difference in this world. With the dead one you cannot marry, to the dead one you cannot be a friend, you can't talk with him, you don't touch him. If you march West, soldier, we say good-by, we who are alive.

Men and women, dogs and birds and insects—they shall not be with you any more. . . .

Soldier, farewell.

Today, you are one of us; you are one with men and women and everything that lives. You are master of the rocks and the woods and all inanimate things.

Tomorrow you march West.

Soldier, farewell.

Do you hear our voice?

Farewell.

Quoted by Hans Speier, "Morale and Propaganda," in Hans Speier and Alfred Kähler (eds.), *War in Our Time* (New York: W. W. Norton & Co., Inc., 1939), Ch. XV, pp. 322-23.

Index